ECOLOGY
A Bridge Between Science and Society

ECOLOGY

A Bridge Between Science and Society

Eugene P. Odum

Institute of Ecology
The University of Georgia

Sinauer Associates, Inc.
Publishers
Sunderland, Massachusetts 01375 U.S.A.

THE COVER
Marshes on Sapelo Island, Georgia. Original watercolor by
Martha H. Odum.

ECOLOGY: A BRIDGE BETWEEN SCIENCE AND SOCIETY
Title in previous editions: *Ecology and Our Endangered Life-Support Systems*.
Copyright © 1997 by Sinauer Associates, Inc. All rights reserved. This book
may not be reproduced in any form without permission from the publisher.
For information contact Sinauer Associates, Inc., 23 Plumtree Road, Sunderland,
MA 01375 U.S.A.

FAX: 413-549-1118
email: publish@sinauer.com

Library of Congress Cataloging-in-Publication Data

Odum, Eugene Pleasants, 1913–
        Ecology: a bridge between science and society / Eugene P. Odum
            p.   cm.
        Rev. ed. of: Ecology and our endangered life-support systems.
        2nd ed. c1993.
        Includes bibliographical references and index.
        ISBN 0-87893-630-0 (pbk.)
        1. Ecology.   I. Odum, Eugene Pleasants, 1913– Ecology and our
endangered life-support systems. II. Title.
        QH541.0293   1996
        577—dc21   96-38031
                                                                                              CIP

Printed in Canada
5 4 3 2 1

This book is lovingly dedicated to my late wife, Martha Ann,
and our late son, William Eugene,
who was also an ecologist.

Brief Contents

Contents

Chapter 8   **Major Ecosystem Types of the World 247**

Epilogue   **The Transition from Youth to Maturity 297**

Preface

In the seven years since the publication of the first edition of this book, ecology has continued to emerge from its roots in biology to become more and more a separate discipline that integrates the study of organisms, the physical environment, and human society, in keeping with the Greek root of the word *ecology: oikos*, the "study of the household," the total environment in which we live. Ecology, in my opinion, has now matured enough to be viewed as the basic science of the total environment. More and more colleges and universities are forming or expanding ecological and environmental institutes, centers, and schools that bring together the biological, physical, and human aspects of our world. Environmental courses are offered at all levels of education, from grade school through college.

Especially significant is the rise of "interface" fields of study, with their new societies, journals, symposium volumes, books—and new jobs. Some examples include ecology–economics; ecology–engineering (restoration ecology); ecology–wildlife management (conservation ecology); ecology–taxonomy (biodiversity); ecology–chemistry (ecotoxicology); ecology–geography (landscape ecology); ecology–hydrology (wetland ecology); ecology–agriculture (agroecology); and ecology–philosophy (environmental ethics). If you are considering a career in an environmental area, my advice is to get into one of these interface fields where the real-world problems lie. Because ecology is an integrative science, it has tremendous potential to provide a communication bridge between science and society—a concept emphasized by the new subtitle for this edition. The concept of our endangered life-support systems, the subtitle for the first and second editions, remains an important unifying theme of the text.

In 1959, the English author C. P. Snow wrote a widely cited book entitled *The Two Cultures*, in which he expressed concern about the lack of communication between the sciences and the humanities in academe. In a second edition, published in 1963, he suggested that a "third culture" would have to emerge in order to close this communication gap, which seems to be getting wider.* I venture to suggest that ecology as presented in this book is a candidate for this "third culture," linking not only the natural and social sciences but, more broadly, science and society.

In this new edition, I have focused especially on updating those principles that relate directly or indirectly to current environmental problems. For the Suggested Readings list at the end of each chapter I have tried to select from the huge outpouring of environmental literature of the 1990–1995 period those references that best summarize new concepts and approaches.

As stated in the preface to the First Edition, this book is intended to be not only a textbook for beginning students, but also a citizen's guide to the principles of modern ecology as they relate to today's threats to our earth home. Also, I have kept in mind specialists from fields such as the humanities and social sciences, engineering, agriculture, forestry, public health, law, politics, and economics who are increasingly involved in environmental matters and who need a review of the major principles of ecology in order to broaden their expertise.

Overall, the book can be considered a guide to human ecology because the relevance of the principles discussed to human affairs is stressed throughout. For this reason the third edition may be especially appropriate for environmental literacy and other courses in which nonscience as well as science majors are enrolled. My emphasis is on causes of and long-term solutions to our environmental problems rather than the "quick-fix" treatment of symptoms that has too often been our approach. The energetic interrelationship between natural, agricultural, and urban ecosystems and the need to shift attention from the output of production systems to the management of inputs in order to reduce pollution are especially emphasized.

For a general reading (not associated with a course), I suggest you start with the Prologue and the first three chapters, then shift to the Epilogue, where applications are discussed. If you want to go further, explore the middle chapters for more technical information, documentation of concepts, examples, and discussion of the work of some of the pioneers in ecological thought. Also, I would like to call your attention to the commentary boxes scattered throughout the text; I had a lot of fun writing these sometimes provocative paragraphs!

EUGENE P. ODUM
December 1996

---

*Snow, C. P. 1963. *The Two Cultures: A Second Look.* Cambridge University Press, New York.

Prologue

The Flight of Apollo 13

If there had been anyone on the Fra Mauro region of the moon on April 14, 1970, watching for the arrival of the spacecraft Apollo 13 from the earth, they would have had a long wait. Scheduled to touch down at 7:00 P.M. Eastern Standard Time, the lunar landing craft of Apollo 13 never arrived, because just as the spacecraft was approaching the moon, an explosion knocked out its main life-support system. The lunar module then had to be pressed into service as a "lifeboat" in order to return the astronauts safely to earth. The dramatic and suspenseful return flight gripped the world's attention for the three days required to get the three astronauts back from lifeless space to life-giving Mother Earth. All nations put aside their troubles and conflicts and offered assistance and prayers. For this brief period there was truly a one-world atmosphere, because when life support is threatened there can be only one mission—survival.

The story of Apollo 13 has been made into a movie that topped box office charts in 1994, and the mission commander, Captain James A. Lovell, has published his account (*Lost Moon: The Perilous Voyage of Apollo 13* by James Lovell and J. Kluger). The story of the ill-fated mission is worth remembering and retelling not only as an example of human heroism and ingenuity in pulling back from the brink of disaster, but also for its relevance to our predicament here on "Spaceship Earth." Our global life-support system that provides air, water, food, and power is being stressed by pollution, poor management, and population pressure. It is time to heed the early warning signs that are appearing: for example, excessive erosion of our best agricultural soils, dying trees in industrial regions, declining ocean fisheries, and the increase in environmentally related human health problems.

The Countdown

The National Aeronautics and Space Administration (NASA) planned a ten-day mission for Apollo 13: three days out, three days back, and four days orbiting the moon, during which time the lunar landing craft would descend for a soft landing on the surface for a 33-hour stay. During that time, two walks lasting 4 or 5 hours each were planned for exploration, collecting rocks (about 95 pounds were to be brought back to earth), and setting up instruments. For the first time, a power drill was to be used to

You Can't See the Picture When You're Standing Inside the Frame

From time to time we neglect some aspect of our well-being because preoccupation with the problem of the moment obscures the big picture. Then some crisis or spectacular event brings on a sudden awakening and a rush to correct the oversight. What can best be described as a worldwide environmental awareness movement burst on the scene during the two-year period of 1968–1970, when astronauts went into space and brought back the first pictures of the earth from afar. For the first time in history, we were able to get outside the frame and see the whole earth! As a result, an annual Earth Day was launched in 1970, and many environmental protection laws were enacted in the United States and in other countries. Public opinion forced governments and industries to consider the possible detrimental environmental impacts of proposed uses and alterations of land and water resources (the environmental impact assessment period). Then, in the 1980s, environmental issues were pushed into the political background by concerns for human relations problems such as crime, the Cold War, governmental budgets, welfare, and so on. Now, as we approach the twenty-first century, environmental concerns are again coming to the forefront. Let us hope that this time, to use a medical analogy, our emphasis will be on prevention rather than treatment.

take cores from the rocky surface. Also for the first time, color television pictures were to be sent back to earth. The landing was planned for the Fra Mauro region, named after a fifteenth-century monk, geographer, and cartographer. The site is a low, rolling Piedmont-like terrain with numerous craters. Rocks at this site are thought to be the oldest on the moon, perhaps dating back to the formation of that lifeless body.

On April 11, 1970, Apollo 13, the fifth in the series of Apollo moon missions, was launched from Cape Canaveral, Florida. The spring of the year before, Neil Armstrong had uttered his historic proclamation, "One small step for man, one giant leap for mankind," as he stepped out of the Apollo 11 lunar landing module to become the first human to walk on the moon. And in the autumn of 1969, Apollo 12 made a second successful landing on the moon. Pictures of the earth taken from the moon during these landings showed us how unique and beautiful our planet was and how fragile and alone it looked in space (Figure 1).

FIGURE 1
The earth as seen from the moon. The earth is a watery planet, with extensive oceans (dark areas) and cloud cover, but 20% of its land areas are deserts, many of which are spreading due to human mismanagement. (Photograph courtesy of NASA.)

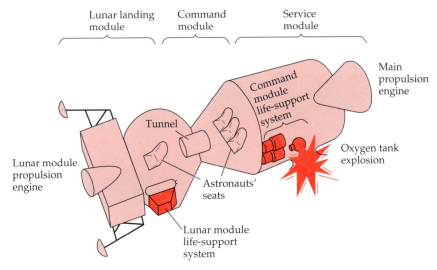

FIGURE 2

Apollo 13. When the oxygen tank exploded, knocking out the life-support system for the command module, the three astronauts had to crowd into the lunar landing module, which had barely enough life-support "consumables" to get the astronauts back to earth.

As shown in Figure 2, the spacecraft consisted of three modules: (1) the service module containing large rocket engines, fuel cells, and other life-support apparatus that provided power, oxygen, and water; (2) the command module, code-named *Odyssey*, the home for the astronauts; and (3) the lunar landing module, code-named *Aquarius*, which would separate from the command module for the short trip down to the moon surface and back.

Three astronauts were in the command module as Apollo 13 began its journey that fine morning in Florida: Captain James A. Lovell, mission commander; Fred W. Haise, Jr., lunar module pilot; and John L. Swigert, Jr., command module pilot. Swigert was a last-minute replacement for Lieutenant Commander T. K. Mattingly, who was exposed to German measles (raising the possibility that he might come down with the disease during the flight).

The countdown and launch of Apollo 13 were picture-perfect. Captain Lovell's first report as he prepared to leave the earth's orbit was: "It looks good to be up here again." (He had circumnavigated the moon once before on Apollo 8.) For the next two days, the flight continued to go well—so routinely, in fact, that the world and the news media lost interest and turned to other events and news. On the evening of the second day, after beaming back a televised tour of the spacecraft, Lovell ended the show by saying, "The crew of Apollo 13 wishes everyone a nice evening. We are ready to close out our inspection and get back to a pleasant evening in *Odyssey*."

The Explosion

Suddenly, at 10:08 P.M. on April 13, as the spacecraft neared the moon, an explosion occurred in the service module and alarm lights flashed on the command module control panel. Swigert's voice crackled out sharply, "Houston, we've had a problem here." At the Houston Manned Space Center, Cap Com (Capsule Communicator) Jack Lousma replied, "Say again, please." Captain Lovell replied, "We've had a main B-bus under-volt." (Translated, there had been a power loss in one of the two main electrical harnesses, called "buses.") He continued, "We had a pretty large bang associated with the warning."

Then the pressure in one of the two oxygen tanks in the service module plummeted to zero, and pressure in the other tank began to drop. It was quickly evident that the explosion had ruptured one or both oxygen tanks. The astronauts could see the precious gas being vented from the side of the service module. Two of the three fuel cells that required oxygen to produce electricity were also rapidly failing.

At this point, all thought of the moon mission was abandoned. Computers and staff at Mission Control sprang into action to plot rescue options that would use the lunar module, which had its own life-support apparatus, as a lifeboat. Thus, at shortly after midnight, there was mounted the most massive and far-flung rescue operation in history, involving a thousand or more individuals at Mission Control and many thousands more on ships in the Pacific Ocean, where the astronauts would eventually have to come down.

It was not immediately known how long the command module would be livable, because information on its "consumables" was in pieces, with no total picture on which to base an estimate of how much time was available to get the astronauts safely back to earth. Precious time was wasted getting all the pieces of information together. (One is reminded here that a similar situation exists here on earth: we do not have a total picture of our life-support "consumables" or understand how they interact, and we don't know how much time we would have on earth should a nuclear "large bang" occur.)

Lovell and Haise entered the cabin of *Aquarius* and switched on the lunar module's independent electrical and oxygen systems. Swigert remained in the command module, breathing oxygen from the lunar module through a hose cannibalized from a space suit. A makeshift extension cord was rigged up to bring power in from the lunar module. Fortunately, it was possible to swing the spacecraft around to the back side of the moon and direct it back to earth using the lunar module's rocket engines, rather than risk firing the main engine in the service module, which might have been damaged by the explosion. All four "burns" necessary to do this came off without a hitch.

The astronauts prepared to conserve as much as possible of their meager supplies of oxygen and power during the three-day return trip. It was an uncomfortable trip, since the temperature of the cabin was allowed to go down almost to freezing. Carbon dioxide began building up to dangerous levels in the lunar module, since the canisters of lithium hydroxide used to absorb it were designed only for the limited time that was to be spent on the moon. A jury-rigged hose was connected to the lithium hydroxide canisters in the command module.

When the spacecraft reached the earth's atmosphere, there was enough power left to charge batteries inside *Odyssey* so it could be reoccupied. The service module and the lunar module were then successfully jettisoned, leaving the command module to float down to the Pacific Ocean where ships were waiting.

In the aftermath, the superstitious among us wondered why NASA chose the number 13 for the mission (the trouble began on April 13, but it was a Monday, not a Friday). Since the service module did not return to earth, the cause of the explosion could not be determined. NASA announced that the probable cause was a short circuit, either in a fan inside one of the oxygen tanks or in the wiring leading to it. An immediate redesign was ordered—the fans were removed and the wiring was changed. There was no further trouble with the tanks in the subsequent five moon flights (Apollo 18 was the last).

Some other problems encountered during the return trip came to light. Leaking water got the astronauts' feet wet, and there was the problem of what to do with accumulated urine in the overcrowded lunar module. (This reminds us of the problem of sewage disposal in an overcrowded city.) Since dumping the urine into space might cause the course of the spacecraft to be altered, it was stored in bags that happened to be on board. There was also the problem of what to do with the eight pounds of plutonium that was to have been left on the moon to power experimental equipment. (Again we are reminded of the unsolved problem of what to do with radioactive wastes here on earth.) It was finally decided to dump it into the Pacific when the lunar module was jettisoned. So somewhere in the ocean depths lies a radioactive memento of ill-fated Apollo 13.

The Contrast between Spacecraft and Earth Life-Support Systems

The life-support systems used so far in manned space flights have been mechanically controlled "storage systems." For the most part, vital necessities such as oxygen and food are produced on earth and stored on board, not regenerated as they are here on earth. Likewise, waste products such as carbon dioxide and urine are chemically stored, not recycled. In contrast, the earth is **bioregenerative:** plants, animals, and especially microorganisms regenerate, recycle, and control life's necessities. Since we did not build the earth's life-support systems, and since they involve a complex array of subsystems, we do not have a clear understanding of how the whole thing works. So far, all attempts to build a large bioregenerative life-support system that would support a large number of people in space, without a supply "umbilical cord" to earth, have failed. At this point, our stay in space is limited by the amount of life-supporting "consumables" that can be carried on board or replenished by supply flights from earth. A number of designs for a regenerative life-support system are currently being tested, including Biosphere-2, which is pictured and described in Chapter 1. Prototypes that provide for some recycling are being developed for use in space stations by space agencies in the United States and Russia.

We need to learn a lot more about how the current real-world life-support systems of our earth function, not only so we can preserve and maintain the quality of these systems, but also so that we might someday build self-maintaining spacecraft and consider colonizing other planets or establishing space colonies on a large scale. More important is the need to understand how the life-supporting nonmarket (unpriced and taken for granted) goods and services provided by the natural environment support and interact with economic, social, cultural, and most other human endeavors. In a very broad sense, the science of ecology provides the background for this understanding.

The chapters that follow will present a broad overview of the earth's vital processes in a form that is intended to be both interesting and understandable. The first three chapters present the "big picture" of life on Spaceship Earth in nontechnical language. The next five chapters provide technical details and "down-to-earth" examples. The Epilogue reviews what we can learn from ecology that can help us to understand and deal with human predicaments.

1

The Life-Support Environment

According to our best understanding of geological history, the earth did not in the beginning support life. The first tiny microorganisms that appeared more than two billion years ago had to survive in an environment with no oxygen, lethal ultraviolet radiation, poisonous gases, and extreme temperature variation, conditions that would be lethal to much of today's life. Over millions of years, organisms (beginning with the first green cyanobacteria) interacting with geological and chemical processes gradually changed the environment. They released oxygen into the atmosphere (as a by-product of photosynthesis), reduced the carbon dioxide level (through limestone formation, especially by marine organisms), and formed a green mantle over the surface of the earth, where sunlight could be converted into all manner of foods to support an increasing number and variety of creatures, including, eventually, humans. (Chapters 3 and 7 will go into more detail on this incredible process, which has continued despite periodic geological setbacks and mass extinctions.) We are able to breathe, drink, and eat in comfort because millions of organisms and hundreds of processes are operating to maintain a livable environment. But we tend to take nature's services for granted because we don't pay money for most of them. Contrary to what some economists say, there are no substitutes for most of nature's resources—water, for example—should they become reduced or deteriorated.

Since life support is provided by a vast, diffuse network of processes operating on different time scales and without set-point controls (there is no global thermostat, for example), we cannot just go out into the environment and point and say: "Look, there is our life-support system ticking away," the way we might point to an air conditioner in our home or the life-support module in a spacecraft. As the old proverb goes, "Out of sight, out of mind"—at least until trouble develops somewhere. Yet the ecological systems and processes that provide life support can be identified. To do this we must think about our environment as a whole and partition the landscape into functional units in some systematic manner.

If we take an airplane trip from Chicago to Europe, we will be looking down for most of the time on large bodies of water: the Atlantic Ocean, lakes, rivers, bays, and so on. These are a very important part of earth's life-support module, since they provide water and function as air purifiers, temperature moderators, and waste assimilators. And water covers more than two-thirds of the earth's surface. Flying over land, we may see large stretches of the same kind of habitat—farmland, grassland, forest—but where humans are concentrated, the landscape is "patchy," with fields, woodlands, towns, cities, suburbs, and highways, often arranged in seemingly haphazard fashion. What we see in an aerial view can be listed under three categories following a classification often used by students and professionals in the field of landscape design: **fabricated**, **domesticated**, and **natural environments**. In less formal language we can think of the landscape as being divided into **developed sites**, **cultivated sites**, and **natural sites**.

The fabricated or developed environment includes cities, industrial parks, and transportation corridors such as roads, railways, and airports. From the standpoint of energy use, we can think of the fabricated environment as comprising fuel-powered systems. For much of the world today the fuel that runs our great cities and industries is fossil fuel—coal, oil, and natural gas—natural products, strange as it may seem, produced

FIGURE 1
Examples of the two major life-supporting landscape types. (A) A well-managed agricultural landscape in Iowa with a strip planting of grass and corn on sloping land. (Photograph courtesy of Soil Conservation Service.) (B) A natural hardwood stand in the Pisgah National Forest in North Carolina. The picture was taken at the entrance to a public campground. (Photograph courtesy of U.S. Forest Service.) The nonmarket value of the natural landscape is equal to or greater than the market value of the domesticated landscape; both are vital to the continued well-being of human societies.

(A)

(B)

in bygone geological ages. Urban–industrial developments actually cover a small area of our total landscape, but they are so energy intensive—that is, they require so much energy and create so much waste heat and pollution—that they have an enormous impact on the other two environments. For example, the **energy density** (the amount of energy consumed per unit of area per year) of an urban–industrial area is 1000 or more times greater than that of a forest. Not only does the city pour its waste products into the countryside, but it depends on this same countryside to provide almost all of its life-supporting resources. Urban–industrial areas are "hot spots" that are supported by very large areas of less energetic environment. Interestingly enough, humans are not the only organisms that aggregate in "hot spots" (see box on page 62).

The domesticated environment includes agricultural lands such as the farm in Figure 1A, managed woodlands and forests, and artificial ponds and lakes. Cultured plants and domestic animals dominate this environment, which is modified and managed so as to promote production of food and fiber, recreation, and other human uses. This part of our landscape is made up of what ecologists often call **subsidized solar-powered systems**. The sun provides the basic energy, but this source is augmented by human-controlled work energy in the form of human labor, machines, fertilizers, and so on, much of which is derived from fuels. Parts of this environment, such as industrialized farms, are quite energy intensive and have considerable impact on the other two environment types due to water, soil, fertilizer, and pesticide runoffs.

"Self-supporting" and "self-maintaining" are the key words characterizing the natural environment. Natural areas, like the forest in Figure 1B, operate without energetic or economic flows directly controlled by humans. These are the **basic solar-powered systems,** dependent on sunlight and on other natural forces, such as rainfall, water flow, and winds, which are indirect forms of solar energy. Also, gravity is involved in the movement of water and other materials. Natural environments include not just the wilderness where few people go, but many sites that are familiar to all of us, such as natural streams, rivers, woodlands, prairies, mountains, lakes, and the oceans. Being self-maintaining does not mean that the natural environment is not used or affected by humans. A national forest, for example, may be grazed by sheep or selectively logged. As long as these uses do not appreciably change the structure and function of the forest or its ability to reproduce itself, then the forest qualifies as a natural area according to our definition. In contrast, a pine plantation with trees planted in rows and harvested all at once in short rotation under strict human management is not a natural area but a cultivated one, like a field of corn. Figure 2 compares a natural forest and a cultivated forest.

(A)

(B)

FIGURE 2
(A) A natural young pine forest in Arkansas developing on abandoned farmland. (B) A domesticated pine plantation, completely lacking in undergrowth and diversity in general. Development in the natural ecosystem is self-organized, without energy input or control by humans. (Photographs courtesy of U.S. Forest Service.)

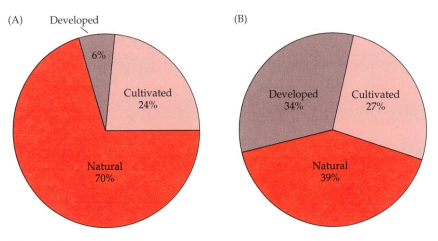

FIGURE 3
Land use in the continental United States as of 1980. (A) Area occupied by each of the three landscape categories. (B) Land use weighted by energy density. Cultivated environments are partitioned to have an estimated energy density twice as high as natural environments; developed environments, ten times as high. Since 1980 the developed segment has been increasing at the expense of both the natural and cultivated areas, putting what had been a reasonable energetic balance in jeopardy.

Figure 3A is a pie chart showing how much of the United States is occupied by each of the three landscape types. The developed environment occupies only a small percentage of the total land area (and an even smaller percentage if we include the surrounding oceans), but its importance in the overall picture is far greater than its area might indicate because of its high energy use, as Figure 3B shows. Figure 4, a nighttime satellite photograph, shows how cities and other highly developed areas dominate the countryside in much of the East, the Great Lakes region, and the West Coast.

We can now define our concept of life-support environment in more precise terms. **Life-support environment** is that part of the earth that provides the physiological necessities of life, namely, food and other energy, mineral nutrients, air, and water. We will use **life-support system** as the functional term for the environment, organisms, **processes,** and resources interacting to provide these physical necessities. By processes we mean operations such as food production, water recycling, waste assimilation, air purification, and so on. Some of these processes are organized and controlled by humans, but many are natural and driven by solar or other natural energies. All life-supporting processes involve the activities of organisms other than humans—plants, animals, and microbes.)

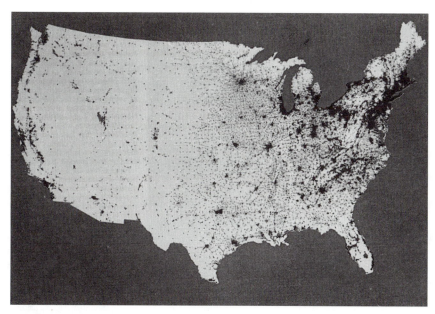

FIGURE 4
The urbanization of America. The dark areas represent light visible at night by satellite. The area that has a population density of 50 or more persons per square mile, and thus a high rate of energy use, is rapidly increasing.

In terms of the landscape, **agricultural systems + natural systems = life-support systems.** Agricultural systems provide the one million calories, and the 15 percent of these calories in protein, that each person requires each year (admittedly, large numbers of people currently are not getting an adequate diet). The natural systems, as already noted, provide the other physiological necessities of life. The word *system* (dictionary definition: regularly interacting items forming a unified whole) is the appropriate term, since life support involves not just an area, but plants, animals, and microbes interacting with water, soil, minerals, atmosphere, and so forth.

The Biosphere-2 Experiment

In 1992, construction was completed on an earthbound capsule designed to be, in part at least, bioregenerative. Called Biosphere-2 (Biosphere-1 being the earth), it is shown in Figure 5. The aerial view shows the 1.27 hectare (3.24 acre), airtight, glass-covered enclosure (labeled 1–5 in the photo) and the external supporting structures (labeled 6–9): the "lungs" that maintain air pressure, the energy center that provides electricity and natural gas for air and water circulation and utilities, and the cooling towers that dissipate the heat that accumulates in the big greenhouse. In the fall of 1991 eight people were sealed in the capsule, where they lived for two years without material exchange with the outside, although with abundant energy flow (as would be required for any life-support system) and unrestricted information exchange (radio, television, telephone, etc.).

FIGURE 5
An aerial photograph of Biosphere-2, an experimental bioregenerative capsule, and its supporting structures. The three-acre, glass-enclosed space (1–5 in photo) combines natural and artificial systems and controls. For two years, 1991–1993, eight people lived in isolation in the capsule with energy inputs (solar and fossil fuels) and information exchanges, but no gaseous, water, or other material exchanges with the outside. The greenhouse enclosure contains life-support environments including rain forest (1), savanna/ocean/marsh (2), desert (3), intensive agriculture (4), and human habitat (5). Also shown are the "lungs" (6 and 7), energy center (8), and cooling towers (9). (Photograph courtesy of Space Biosphere Ventures.)

About 80 percent of the space inside Biosphere-2 is occupied by half a dozen natural habitats, ranging from rain forest to desert. These provide a great deal of biodiversity since it is expected that some of the species in the enclosure will thrive while others may not adapt and will be lost. Most of the rest of the area, about 16 percent, or 0.2 hectares (0.5 acre), is occupied by crops (the agricultural wing), which feed the humans and the small herd of domestic animals (goats, pigs, and chickens) that provides milk and a little bit of meat for the low-cholesterol diet of the human inhabitants. The human habitat where the eight people have their rooms—the urban area, as it were—is very small, about 4 percent of the space. The allotment of space among the three basic environment types we have been discussing in this chapter (natural, cultivated, and developed) is similar to the proportions of land use in the United States as shown in Figure 3A. But in Biosphere-2 there are no automobiles or polluting industries in the "developed area." If there were, or if the population were to increase, a lot more life-supporting environment would be needed. (For additional descriptions and pictures, see Allen 1991 and Alling and Nelson 1993.)

In the fall of 1993 the eight "biospherians" emerged from their two-year isolation still speaking to one another and in better health than when they went in (Walford 1994). The complex machinery that maintained air and water circulation and recycling, heating and cooling, and so on worked quite well. Incoming solar energy was sufficient to maintain the labor-intensive food gardens, including banana plants in tubs located in sunny spots throughout the enclosure. However, total photosynthesis was not quite enough to maintain the oxygen–carbon dioxide balance; during the last six months oxygen had to be added to prevent "altitude sickness." Apparently, reduction of light by the glass, unusually cloudy weather outside, and the rich organic soil brought in for the agricultural wing combined to reduce production and increase consumption of oxygen more than anticipated (Severinghaus et al. 1994).

Some scientists criticized the Biosphere-2 experiment as not being "real science" because the crew were not scientists, but persons who were selected for their ability to work together, grow all their food, man the pumps and control the devices, and for their willingness to live at a subsistence level for two years. The crew, for example, had to spend about 45 percent of their waking hours on growing and preparing food, 25 percent on maintenance and repairs, 20 percent on communication, and 5 percent on small research projects, which left little time for relaxation and recreation. Few scientists could qualify for such a job! Viewed as an exercise in human ecology and environmental engineering, the experiment was a great success.

The High Cost of Life Support

Perhaps the most important result of the first Biosphere-2 experiment was the documentation of the high cost of providing what are largely the free services of Nature with the use of nonrenewable fossil fuels. To maintain life support in the enclosure, electricity was consumed at a rate of 700 kilowatts and natural gas at a rate of 23 million kilojoules per hour (Dempster 1993). In the units that utilities use to calculate your bills, these rates add up to 12.5 million kilowatt-hours (kWh) of electricity and 3.8 million therms of gas for the two-year period. At 10 cents per kWh and 70 cents per therm (average rates for the United States), the costs would be $1.25 million and $2.6 million respectively, or approaching $4 million. If the crew of eight persons had to pay for this energy at residential prices, their monthly bills would be more than $150,000. At these rates, very few of the future 10–20 billion people on earth could afford to live in domed cities (see Odum 1996).

Although Biosphere-2 is closed to material exchange, including water and other liquids, it is open to energy (sun and fossil fuels, in this case) as is necessary in any sustainable life-support system. In future experiments it may be possible to use solar cells (photovoltaics) to produce enough electricity to power the pumps and machines; then the whole thing might run on solar power, as does Biosphere-1 (the earth), and as might projected future space colonies.

In the meantime, the National Aeronautics and Space Administration (NASA) continues to design and experiment with much smaller compact life-support modules for use in space stations or lunar bases. A recent design called Controlled Ecological Life-Support System (CELSS) utilizes hybrid physicochemical/bioregenerative technology—that is, air, water, and food are produced or regenerated by physical, chemical, or biological means, whichever minimizes mass and power requirements (Schwartzkopf 1992).

Supporting Urban–Industrial Areas

Since the area on Planet Earth devoted to agriculture is several times greater than the urban–industrial area (Figure 3A), we can see that many square miles of agricultural land are required to feed the thousand or more people living on each urban square mile. We need to worry about the encroachment of urban development on prime farmland because, contrary to what many people believe, good farmland is limited. Worldwide, only about a quarter of the land area has the soils, water, and climate to sustain the high level of food production needed to feed the earth's billions of inhabitants.

Densely populated countries with a small land area, such as Belgium, Israel, and Japan, depend heavily on areas outside the country for life-supporting necessities. For example, Japanese fishing fleets roam widely in the Pacific and Atlantic Oceans in order to obtain the protein required by Japan's large population, and only about 30 percent of Japan's agricultural food needs comes from land within the country. Georg Borgstrom, who writes books with titles such as *The Hungry Planet* (1967), uses the term **ghost acreage** to denote the unbounded area outside a country required to sustain the population within the boundary of the country. Japan's ghost acreage is significantly larger than its territorial land and adjacent shallow seas. In contrast, the United States, with its much larger land area and much less dense human population, is more than self-sufficient as far as food is concerned, although not when it comes to energy and many vital minerals.

As shown in Figure 3A, a large area of natural environment in the United States is available for providing other vital necessities. Again, because of the heavy demands of the cities, a great many square miles of natural area are needed to support them. Just what would be an optimum ratio between natural and developed lands is difficult to estimate, not only because it depends on the energy density of the densely populated areas, but also because it is extremely difficult to quantify the "goods and services of nature" and to determine just how many life-support resources are "wasted" when cities fail to conserve, clean, and recycle air, water, garbage, and so forth.

Suffice it to say for now that a large area of natural environment is a necessary part of humankind's total environment (Odum and Odum 1972). All that seemingly "vacant" land and water you see looking down from an airplane is not wasteland; it's working continuously day and night to keep you and all the other animals and plants alive and healthy.

Since the United States, and much of the rest of the world, is becoming increasingly urban, it is important that we recognize that *the city is a parasite on the natural and domesticated environments,* since it makes no food, cleans no air, and cleans very little water to a point where it could be reused. The larger the city, the greater the need for undeveloped or lightly developed countryside to provide the necessary host for the urban parasite. When we discuss host–parasite relationships later, we will note that a parasite does not live for very long if it kills or damages its host. The well-adapted parasite not only does not destroy its host—it actually develops exchanges or "feedbacks" that benefit both itself and its host so that both may thrive. And so it must be for the well-adapted, sustainable city of the future.

Wealth produced by the city flows out into the countryside in exchange for the natural and human-made goods and services that flow into the city. Some of this urban-produced wealth needs to be used to preserve, service, and repair the natural and agricultural environment if the quality of city life is to be sustained. At present we do not do an adequate job of caring for our life-support environment, because we don't realize how vital it is. To illustrate, let us now examine the dependence of two cities, New York and Chicago, on their "downstream watersheds."

The New York Bight

Wastes produced by the 20 million people living in the New York area are discharged into the embayment at the mouth of the Hudson River that is bounded by Long Island and New Jersey, as shown in Figure 6. Because of the coastal indentation at this point, the area is known as the New York Bight. More than 10 million tons of solid wastes are dumped here each year, plus unmeasured gallons of treated sewage, industrial wastes, street runoffs, shipping discharges, and so on. Because the large body of water is so physically and biologically active (due to vigorous currents and tides, intense bacterial activity, and so on), it has been able, so far, to "digest" all or most of this enormous discharge.

However, increasing signs of stress, such as beach contamination and fish kills, indicate that the capacity of this life-support environment to assimilate wastes is being exceeded. Hypodermic needles and other hospital wastes have been washed up on beaches. Recent studies funded by the National Oceanic and Atmospheric Administration (NOAA) have shown that the more dangerous residues (pesticides, lead, and a whole host of other poisonous materials) are absorbed by fine-particle sediments, which slowly work their way shoreward into the intracoastal zone's marshes, estuaries, and lagoons (Young et al. 1985). These zones have a high assimilative capacity for many residues, but many of them are being lost to development. There is constant economic pressure to fill in and develop coastal marshes and estuaries, and developers and people in general are unaware of the tremendous value of these areas in their natural state. Waste processing is only one of the life-support services estuaries and wetlands provide for us; others are discussed in Chapter 8.

The 2000 or so square mile (5200 km^2) body of water, with an average depth of about 100 feet (30 m), that comprises the New York Bight functions as a large activated waste treatment system that operates free of

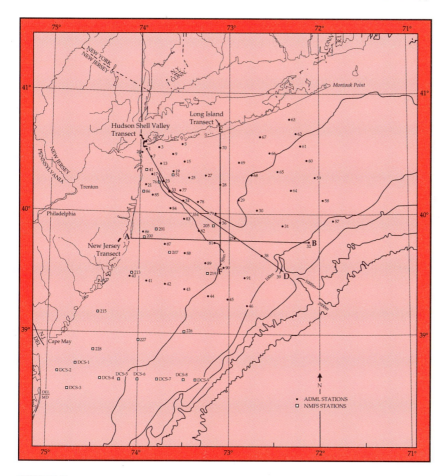

FIGURE 6
The New York Bight, which has become a vast natural waste management system for New York City and the adjacent urban areas. Contours and water quality sampling stations are shown. (From Young et al. 1985.)

charge 24 hours a day, 365 days a year. Think what it would cost to process all that waste in a human-made mechanical treatment system that required expensive fuel instead of the sun and tidal energy that drives the natural system. Or think how much valuable land would be lost to other uses if all that stuff were put in a giant landfill (New York recently passed a law prohibiting the development of new landfills, but this regulation is no solution since wastes have to be exported to less populous regions). New York state and local taxes, already among the highest in the country,

Our Free Sewage Systems

Almost without exception, large cities (population 10 million or more) throughout the world are located on large bodies of water—ocean shores, rivers, lakes, or estuaries—that provide natural waste treatment facilities. See if you can think of any large city that is not so located. Mexico City may come to mind, but air and water quality there are far from being what is desired. Large cities worldwide are having to face up to the fact that their free sewers are becoming overloaded, and won't be free any more when their capacity to assimilate waste is destroyed.

would have to be greatly increased if taxpayers had to pay for the work of the New York Bight. Since, as already noted, this natural waste treatment system has become overloaded in recent years, there are only two options: (1) increase costly artificial treatment, or (2) reduce the amount of waste that requires disposal or treatment. Pollution prevention and waste reduction technology are what is now needed, as we will discuss later in this book.

The Illinois River

For Chicago, the Illinois River basin provides the same vital service the Bight does for New York. In 1900, Chicago decided to divert its sewage away from Lake Michigan and into the Illinois River, which runs southward down to the Mississippi (Figure 7). A canal was dug from Lake Michigan to the headwaters of the river for this purpose (and also to provide a waterway for boat transportation). The Illinois River valley is broad, with hundreds of shallow lakes and sloughs bordering the main channel, which runs through some of the most fertile soils to be found anywhere in the world. Indians and early settlers found the valley to be a game and fish paradise, teeming with waterfowl, fish, mussels, and fur-bearing animals. As late as 1908, 2500 commercial fishermen took 11 million kg of fish from the Illinois River annually, and it was estimated that sport fishermen contributed as much money to the local economies as did the commercial fishermen. There were also 2600 mussel-fishing boats operating during that time. The river was and is an important waterway for transporting grain and fuel.

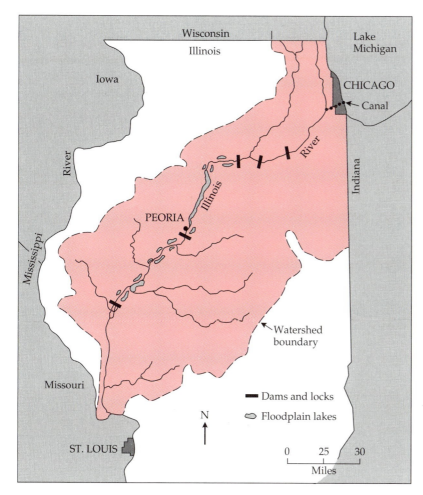

FIGURE 7
The Illinois River drainage basin (enclosed by dashed lines) is increasingly affected by Chicago wastes and nonpoint-source soil and chemical pollution from agriculture. (After Havera and Bellrose 1985.)

The diverted Lake Michigan water briefly increased the spread of the river, but increasing volumes of untreated sewage soon reduced water quality in the upper part of the river, and fish catches declined rapidly. The situation was improved between 1920 and 1930 when cities built sewage treatment plants and water pollution laws were enacted. Locks and dams built during this period had both positive and negative effects on the river basin. In 1940, the river as a whole was in tolerable shape and was assimilating the vast wastes of Chicago and other cities in the valley.

But in the next few decades, a new stress developed that threatens to reduce or even completely destroy the river's capacity as a life-support system. Beginning in the 1930s, the fertile prairie land in the valley was converted to vast row-crop monocultures of corn and soybeans. With economic policies and the availability of cheap fossil fuel and agricultural chemicals encouraging maximum cash crop production, pastures, wood-lots, bottomland lakes, fence rows, and protective vegetation along streams were all converted to grain fields. Soil conservation practices that were well established in the 1930s were abandoned, and yield became almost the sole goal of farming with little regard to how long such yields could be sustained or how much damage was being done to other environments. Soil erosion and runoff of toxic chemicals increased with increasing agricultural production. In 1975, it was estimated that 25 million tons of soil were moving from farmland into the river system each year. Some of this soil goes on to the Mississippi, but most of it settles in the shallow lakes and lagoons that are such a valuable part of the Illinois River basin. Sedimentation and chemical poisons are now the greatest threat to the Illinois River. The dilemma is that efforts to enhance one part of the life-support environment (agriculture) have degraded other equally vital components (natural systems). (For more on the Illinois River story, see Havera and Bellrose 1985. Present efforts to restore the river are described by Sparks 1992.)

Trends similar to those described for the Illinois River are occurring throughout the United States and in other parts of the world as well. San Francisco Bay is threatened by the continued disposal of toxic wastes, and by increased salinity due to runoff from irrigated farmlands and the diversion of fresh water for agriculture and domestic uses upstream (see Nichols et al. 1986). The waters of Chesapeake Bay, one of the world's finest estuaries and home of fabulous oysters and blue crabs, increasingly suffers from oxygen depletion and a general decrease in water quality. Only by considering the whole Chesapeake watershed (120,000 square miles extending from just south of Syracuse, New York, to Norfolk, Virginia, and containing some 15 million people) can the decline in water quality be reversed. Because there is now a consensus that the area is in decline, remedial action can be achieved if (and that is a big if) a coalition of social, economic, and environmental interests can be held together as the painful steps are taken to reduce the inflows of soil, pesticides, heavy metals, and so on. If successful, the Chesapeake project could be "a model for sustainable ecosystem management worthy of emulation" (Constanza and Grier 1995.) Especially hard hit by these problems are the semi-enclosed seas of the Old World with growing human populations on their shores, such as the Baltic, the Mediterranean, the Caspian, the Red, the Black, the South China, the Irish, and the English Channel (Platt 1995).

Point-Source versus Nonpoint-Source Pollution

Passage and enforcement of laws regulating discharges from industry, power plants, and sewage treatment plants have reduced **point-source pollution** in many streams and rivers as well as in the air. Discharges from pipes, ditches, smokestacks, and other sources, like the factory in Figure 8, are relatively easily located and acted upon. However, **nonpoint-source pollution** (such as soil and pesticide runoff from agricultural lands, or auto exhaust) has increased (Smith et al. 1987), so there has been little or no overall improvement in water and air quality—even a decline in the cases described. But take heart; new ideas about how to reduce the stress on our life-support environment are emerging.

Control of point-source pollution of Lake Erie and the other Great Lakes has greatly improved water quality and has restored some fish populations, but water quality is now threatened by industrial and agricultural chemicals coming in via groundwater. The nonpoint-source pollutants that now pose such a serious threat to lakes, rivers, oceans, and the atmosphere are more difficult to assess, and they cannot be controlled, as can point sources, on the "output" side. Nonpoint-source pollution can be controlled only by **input management**; by, for example, reducing the amount and toxicity of agricultural chemicals applied to cropland, removing sulfur and other pollutants from coal before it is burned in power plants, or recycling paper instead of dumping it in a landfill. Since reducing costly inputs can increase profits on agricultural and industrial products, there are strong economic reasons for an about-face on the way we deal with wastes. The landfill and other "dumps" must now be considered obsolete, and must be replaced by waste recovery industries. (The concept of input management is diagrammed and discussed further in the Epilogue.)

The New York Bight and the Illinois River illustrate not only the value of natural environment life-support systems, but also the need to give more attention to increasing the efficiency of resource use, thereby reducing the deleterious impacts of the fabricated and domesticated environments on the life-support environment.

The time has come to view and manage entire landscapes as a whole. And this is where the science of ecology can help, since it deals with the interconnectedness of humans and nature. The word *ecology* is derived

FIGURE 8
While point-source pollution, such as the direct discharge of untreated wastes shown in this 1950 photo, is being reduced in the United States, nonpoint-source pollution, which is not so easily photographed, is on the increase worldwide. (Photograph courtesy of Soil Conservation Service.)

from the Greek *oikos*, meaning "household," combined with the root *logy,* meaning "the study of." Thus, **ecology** is, literally, the study of households, including the plants, animals, microbes, and people that live together as interdependent beings on Spaceship Earth. As already emphasized, the environmental house within which we place our human-made structures and operate our machines provides most of our vital biological

> ## More than Knowledge
>
> The study of ecology gives us more than practical knowledge. It also shows us the earth's fantastic beauty and the incredible variety of life. Despite our increasing dependence on machines and human-made structures, the love of nature remains a powerful force in the human psyche. Aesthetic values and the conservation ethic are deeply rooted, even though too often overshadowed by greed and the pursuit of short-term economic and political gain (which, unfortunately, many of us equate with the pursuit of happiness). The paradox is that although urbanization increasingly separates us from the natural environment, we are increasingly concerned about it because we are increasingly dependent on it.

necessities; hence we can think of ecology as the study of the earth's life-support systems.

The next several chapters will introduce the principles of a holistic ecology in the belief that informed citizens sooner or later will see to it that future "progress" includes sustaining the quality of our total environment. If natural life-support goods and services are accorded equal value with human-made goods and services and managed wisely, quality of life and economic development need not be conflicting goals.

Suggested Readings

*Allen, J. 1991. *Biosphere-2: The Human Experiment*. Penguin Books, London.

*Alling, A. and M. Nelson. 1993. *Life Under Glass: The Inside Story of Biosphere-2*. The Biosphere Press, Oracle, AZ.

*Borgstrom, G. 1967. *The Hungry Planet*. Macmillan, New York. (Concept of "ghost acres," Chapter 5, pp. 70–86.)

Cloud, P. 1988. *Oasis in Space: Earth History from the Beginning*. Norton, New York.

Costanza, R. and J. Greer. 1995. The Chesapeake Bay and its watershed: A model for sustainable ecosystem management? In *Barriers and Bridges*, ed. C. S. Hollings and S. S. Light, pp. 169–213. Columbia University Press, New York.

*Dempster, W. F. 1993. Biosphere-2 systems dynamics during the initial two-year closure trial. Society of Automotive Engineers (SAE) technical paper series, no. 932290. Warrendale, PA.

Dorney, R. S. and P. W. McLellan. 1984. The urban ecosystem: Its spatial structure, its subsystem attributes. *Environment* 16(1): 9–20. (Natural, agro, and urban landscapes are considered as the three basic ecosystems, with the latter two considered as "islands" or "subsystems" in the matrix of natural landscape.)

Echart, P., ed. *Life Support and Biospherics*. Herbert Utz Publishers, Munich.

Ehrlich, A. H. and P. R. Ehrlich. 1987. *Earth*. Franklin Watts, New York.

*Greeson, P. E., J. R. Clark and J. E. Clark, eds. 1979. *Wetland Functions and Values: The State of Our Understanding*. American Water Resources Association, Minneapolis.

*Havera, S. P. and F. C. Bellrose. 1985. The Illinois River: A lesson to be learned. *Wetlands* 4: 29–40.

Hutchinson, G. E., ed. 1970. *The Biosphere*. Special issue, *Sci. Am.* 223(3): 44–208. Also published in book form by W. H. Freeman, San Francisco.

*Nichols, F. H., J. E. Cloern, S. N. Luoma and D. H. Peterson. 1986. The modification of an estuary. *Science* 231: 567–573.

Odum, E. P. 1977. The life support value of forests. In *Forests for People*, pp. 101–105. Society of American Foresters, Washington, D.C.

Odum, E. P. 1996. The high cost of domed cities. *Nature* 382: 18.

Odum, E. P. and E. H. Franz. 1977. Whither the life-support system? In *Growth Without Ecodisasters?* ed. N. Polunin, pp. 264–274. Macmillan Press, London.

*Odum, E. P. and H. T. Odum. 1972. Natural areas as necessary components of man's total environment. *Trans. N.A. Wildl. and Nat. Res. Conf.* 37: 178–189. Wildlife Management Institute, Washington, D.C.

*Platt, A. E. 1995. Dying seas. *Worldwatch* 8(1): 10–19.

*Schwartzkopf, S. H. 1992. Design of a controlled ecological life support system: Regenerative technologies are necessary for implementation in a lunar base CELSS. *BioScience* 42: 526–535.

*Severinghaus, J. P., W. S. Broecher, W. F. Dempster, T. MacCallum and M. Wahlin. 1994. Oxygen loss in Biosphere-2. *Trans. Am. Geophysical Union* 75: 33, 25–37.

*Smith, R. A., R. B. Alexander and M. G. Wolman. 1987. Water-quality trends in the nation's rivers. *Science* 235: 1607–1615. (Between 1974 and 1981 some point-source pollution has declined, but nonpoint-source pollution has increased.)

*Sparks, R. 1992. The Illinois River–floodplain ecosystem. In *Restoration of Aquatic Ecosystems*, ed. J. Cairns, pp. 412–432. National Academy Press, Washington, D.C.

*Young, R. A., D. J. P. Swift, T. L. Clarke, G. R. Harvey and P. R. Betzer. 1985. Dispersal pathways for particle-associated pollutants. *Science* 229: 431–435.

*Indicates references cited in this chapter.

2

Levels of Organization

Chapter 1 presented a bird's-eye view of the earth in terms of its three major environments—developed, cultivated, and natural—with emphasis on the systems that provide support for human civilization. To more fully understand this complex world of ours at ground level, it is helpful to think in terms of levels of organizational hierarchies (Simon 1973; Allen and Starr 1982; O'Neill et al. 1986). A **hierarchy** is defined as an arrangement into a graded series of compartments. Picture, for instance, Chinese boxes, with a box inside of a box inside of a box, and so on. Five examples are shown in Table 1. In these hierarchies the series of levels are arranged from the largest to the smallest, but the order could be reversed if one wished to start with the lowest level of resolution.

We are all more or less familiar with the geographic and the military examples. Early in an elementary course in biology, the student is introduced to the physiological and taxonomic "ladders" that show how the human body is arranged and how organisms are classified, respectively. Another example that is becoming increasingly familiar in this modern age is the computer program, with its routines and subroutines that are activated in an orderly sequence to achieve an overall goal. The geographic, biological, and ecological hierarchies are "nested" in that each level is made up of a group of lower-level units; in contrast, the military and other human-organized hierarchies in governments, corporations, and universities are "non-nested" (since sergeants, for example, are not composed of groups of privates).

Table 1 Examples of Levels of Organizational Hierarchies	
LARGE SCALE	
Geographical & Political	Ecological
World	**Biosphere**
Continent	Biogeographic region
Nation	**Region & Biome**
Region	**Landscape**
State (or Province)	**Ecosystem**
County	Biotic community
Town (or township)	**Population** (species)
Human population (ethnic, etc.)	**Organism**
Individual	

SMALLER SCALE		
Taxonomic	Physiological	Military
Kingdom	Organism	General
Phylum	Organ system	Colonel
Class	Organ	Major
Order	Tissue	Captain
Family	Cell	Lieutenant
Genus	Organelle	Sergeant
Species	Molecule	Private

The Ecological Hierarchy

In ecology, the term **population**, originally coined to denote a group of people, is broadened to include groups of individuals of any species that live together in some designated area. In the singular, a population is a group of organisms of the same, interbreeding species; in the plural, populations may include groups of organisms of different species that are linked by common ancestry or common habitat (e.g., plant populations, bird populations, plankton populations). **Community**, in ecology, is used in the sense of **biotic community** to include all of the populations living in a designated area. The community and the nonliving environment function together as an **ecological system** or **ecosystem**. A parallel term often used in German and Russian literature is *biogeocoenosis*, which translated means "life and earth functioning together."

Referring again to Table 1, groups of ecosystems along with human artifacts make up **landscapes**, which in turn are part of large regional units

No Time to Think

An unnatural decoupling of humans and nature is one of the unfortunate results of high population density and urbanization. A person living in a large city is so preoccupied with surviving in a high-energy, economically competitive environment that the life-supporting natural and agricultural environments are not only out of sight but very much out of mind. Likewise, the millions of poor people in undeveloped countries who must eke out a bare existence on a day-to-day basis have no time or energy to consider the long-range consequences of their actions. To get people started thinking about these things requires that environmental education begin at a very early age. Children often can grasp holistic and commonsense ideas better than adults, who have become specialized and "set in their ways" of thinking.

(both geographical and natural regions) called **biomes** (e.g., an ocean, a grassland region). The major continents and oceans are the **biogeographic regions**, each with its own special flora and fauna. **Biosphere** is the widely used term for all of the earth's ecosystems functioning together on a global scale. The term **ecosphere** is often used as a synonym for *biosphere*, but environmental science textbooks often make a distinction as follows: biosphere = all the life on earth (all the communities); ecosphere = all the life and interacting nonliving materials (all the ecosystems).

All of the levels in the ecological hierarchy involve life and biological processes, so we can think of the biosphere as being that portion of the earth in which organisms can live; that is, the biologically inhabitable soil, air, and water. The biosphere or ecosphere merges imperceptibly (that is, without sharp boundaries) into the **lithosphere** (the rocks, sediments, mantle, and core of the earth), the **hydrosphere** (surface and ground water), and the **atmosphere**, the other major subdivisions of Spaceship Earth.

Order Out of Chaos

Each level in a hierarchy influences what goes on at adjacent levels. Processes at lower levels are often constrained in some way by those at higher levels. For example, interactions at the population level, such as those between a parasite and its host, tend to be nonequilibrium, cyclic, or even chaotic. In contrast, the slower, long-term interactions that characterize higher hierarchical levels tend to approach a **pulsing steady state.** In other words, large ecosystems as a whole, such as oceans or large forests, are less variable over time than their individual components. We might think of this as **hierarchical constraint**. Accordingly, what we see at one level may be different from the situation at another level.

The Importance of Time and Scale

Time and scale are especially important considerations. In the short term, a parasite and its host often engage in an oscillating "arms race" in which each seeks to get the better of the other. But over the long term parasites and hosts tend to adjust to one another and achieve some sort of balanced coexistence. This balance may persist for as long as the adjustments are not disrupted by some large-scale disturbance. (We will learn more about this *coevolution* in Chapter 7.) Likewise, the behavior and survival of a population in a small tract of forest depends on the nature of the large-scale landscape of which the tract is a part. (We will learn more about these *scale effects* in the section on "Source–Sink Energetics" in Chapter 4; see also Meentemeyer and Box 1989.)

The Ecology–Economics Initiative and Other Interfaces

The word **economics** is derived from the same root, *oikos* ("household"), as is the word *ecology*. Since *nomics* means "management," *economics* translates as "the management of the household." In theory, ecology and economics should be companion disciplines. In practice, however, economists deal with human works and with market goods and services, while ecologists have until recently focused on the natural environment and the mostly nonmarket but nonetheless vital goods and services of nature (air purification, water recycling, soil enrichment, etc.). The result of both disciplines taking an overly narrow viewpoint is that the general public tends to view ecologists and economists as adversaries with antithetical visions. More will be said later about the importance of taking a more holistic view of our household, and what is being done to bridge the gap between ecology and economics. In the Epilogue we will discuss the concept of "dual capitalism" as the ultimate bridge.

The realization that most real-world problems require the attention of more than one traditional discipline has resulted in the emergence of many new interface fields, each with its own professional societies, journals, textbooks, and college courses. Ecological economics is one such interface field; others include **environmental health, environmental engi-**

> ## Political Zigzagging
>
> An alternation of individualistic and holistic philosophies seems to be a characteristic feature of politics—in other words, attention shifts back and forth from the individual level to the level of the whole community, nation, or world. Political regimes that promote the "individual good," the so-called conservative stance, tend to alternate with those that focus on the "public good," the so-called liberal philosophy. In the political history of humankind, it has proved difficult to merge the two because they are viewed as competitors. What usually happens is that excessive attention to one level leads to the neglect of the other level, which brings on a new political regime that promises to deal with the neglected level. So, in a sort of zigzag fashion, humanity strives to achieve a balance between what are viewed as individual human rights and public needs (Schlesinger 1986).

neering, **environmental law, conservation biology, landscape ecology**, and **restoration ecology.** The effect that these new specialties will have on future decision making and planning will depend on the soundness of the ecological theory that is applied, the public's understanding of basic ecological principles, and perhaps most of all, how many practical strategies for better management emerge from them.

While almost everyone will agree that awareness of and concern about environmental quality must continue to have a high priority in human affairs, it is difficult to maintain a high level of public interest when there are so many other concerns that demand our attention, even when it can be shown that these other concerns are directly related to ecological concerns. The main point is that while different levels of organization have different, often unique, features, they are all linked together by **transcending functions**, such as energetics, growth, evolution, feedback control, and other common denominators (see Barrett, Peles, and Odum 1997). Accordingly, what happens at any one level affects what happens at another level. And this brings us to the next important ecological principle.

The Emergent Property Principle

An important consequence of hierarchical organization is that as components, or subsets, are combined to produce larger functional wholes, new properties emerge that were not present or not evident at the level below. Accordingly, an **emergent property** of an ecological level or unit is one that results from the functional interaction of the components, and therefore is a property that cannot be predicted from the study of components that are isolated or decoupled from the whole unit (Salt 1979). This principle is a more formal statement of the old adage that the "whole is more than the sum of the parts," or, as is often stated, "the forest is more than a collection of trees." One example from the physical realm and two from the ecological realm will illustrate emergent properties. When oxygen and hydrogen are combined together in a certain molecular configuration, water is formed, a liquid with new properties quite different from those of its gaseous components. When certain algae and coelenterate animals evolve together to produce a coral, an efficient nutrient cycling mechanism results that enables a coral reef to maintain a high rate of productivity in waters with a low nutrient content. Likewise, when certain fungi colonize roots of trees, the resulting fungus–root combinations, known as **mycorrhizae**, are able to extract mineral nutrients from the soil more efficiently than roots alone. Such mutually beneficial relationships are common in nature, and also in a well-ordered human society.

The New Ecology

Ecology is now more and more a discipline that emphasizes a holistic study of both parts and wholes. While the concept of the whole being greater than the sum of the parts is widely recognized, it tends to be overlooked by modern science and technology, which emphasize the detailed study of smaller and smaller units on the theory that specialization is the way to deal with complex matters. In the real world, the truth is that although findings at any one level aid the study of another level, they cannot fully explain the phenomena occurring at that level, which must also be studied to get the complete picture. Thus, to understand and properly manage a forest, we must not only be knowledgeable about trees; we also need to know about the unique characteristics of the forest as it functions in its entirety.

Some attributes, obviously, become more complex and variable as we proceed from small to large units, but as we have already noted, rates of function may become less variable. For example, the rate of photosynthesis of a whole forest or whole cornfield is less variable than that of an individual leaf or plant within the community, because when one leaf, individual, or species slows down, another may speed up in a compensatory manner. More specifically, we can say that **homeostatic mechanisms**, which we may define as checks and balances (or forces and counterforces) that dampen oscillations, operate all along the line. We are familiar with homeostatic mechanisms in the individual as, for example, the regulatory mechanisms in the nervous system that keep our body temperature constant despite fluctuations in the environment. Regulatory mechanisms also operate at higher levels, as, for example, in maintaining the carbon dioxide–oxygen balance in the atmosphere. However, regulation at these higher levels differs in that there are no "set-point" controls such as thermostats or chemostats, as we will see in Chapter 3.

Finally, it is important to recognize that not only do different levels have both different and transcending properties, but also that the effect of an outside disturbance may vary with different levels. For example, consider the vegetation of southern California called *chaparral*, where periodic fires occur every few years during dry seasons (see Figure 16A in Chapter 8). This native vegetation is adapted to fire and could not exist without it. To individual organisms that may be killed or injured by fires, or to a person who has built a home in the chaparral, fire is certainly detrimental. But at the level of the vegetative community, lack of fire would be detrimental. In the absence of periodic fires, the fire-dependent species would be replaced by others, and the entire nature of the vegetation and its animal associates would be changed. (Fire-adapted ecosystems will be discussed further in Chapter 5.) Likewise, flooding on a river floodplain can be a bad thing for an animal caught in rising waters, or for a person who has been unwise enough to build a house on the floodplain, but it is a good and necessary thing for adapted floodplain vegetation. Recent studies have shown that human efforts to suppress so-called "brush fires" in California have reduced the frequency of fires, but have made those fires that do occur more severe, because the suppression of small fires allows fuel (dry, dead wood and leaves) to build up (Minnich 1983). The same may be said for some of our well-intentioned flood control efforts; small floods are controlled, but big floods are worse (Belt 1975).

Insect Pests and Exotic Species

A striking example of the difference between integration and lack of integration of species within their community is seen in cases in which insects become pests when displaced from their native ecosystems. Most agricultural pests are species that live relatively innocuous lives in their native habitat, but become troublesome when they invade, or are inadvertently introduced into, a new region or new agricultural system. Many pests in North America come from other continents (and vice versa), as, for example, the Mediterranean fruit fly, the Japanese beetle, and the European corn borer (the list is very long). In their original habitat, these species function as parts of well-ordered ecosystems in which excesses of reproduction and feeding are controlled as a result of long periods of evolutionary adjustment; in new situations that lack such controls, their populations behave like a cancer that can destroy the whole system before controls can be established. As we shall note in later chapters, one of the prices we pay for high crop yields is the increasing cost of and environmental disruption by chemical controls that replace natural ones. Fortunately, a new technology called **integrated pest management**, which involves coordinating natural and artificial controls, is developing, and shows the potential to reduce this cost (Allen 1980; Murdoch et al. 1985; Stone 1992).

Introduced species often raise havoc not only in agriculture but in natural environments as well. A good example is the zebra mussel, a small clam native to the Caspian Sea region of south central Asia. By attaching to the hulls of ships, it got into the Great Lakes of the United States and Canada. In the 1980s the population exploded, and the clams clogged pipes, reduced populations of plankton (on which it feeds by filtration), the basis for fish food chains, and in general became a pest (Ludansky, McDonald, and Neal 1993; Benson and Boydstun 1995).

One would expect that North American predators would discover this abundant new food source, but it often takes time for such evolutionary adjustments to develop. In the meantime, the only effective predators of zebra mussels seem to be two species of gobies (small fish) that were themselves introduced from the Caspian Sea region. Indeed, there are a number of cases in which introduced pests have been brought under control by introducing a parasite or disease from the pest's original habitat.

The increase in international trade and air traffic has greatly increased the spread of opportunistic plants, animals, and diseases, including some very virulent human diseases. Apparently, invaders are more successful when the ecosystem or host population is stressed. For example, often the first sign of overgrazing in natural grasslands is the appearance of an exotic weed. (For more on the problems caused by invaders, see Mooney and Drake 1986; Bright 1996.)

Applying Hierarchy Theory

The phenomena of hierarchical organization, functional integration, and homeostasis suggest that we can begin the study of ecology at any one of the various levels without having to learn everything there is to know about adjacent levels. The challenge is to recognize the unique characteristics of the level selected and then devise appropriate methods of study and/or action. As shown in Figure 1, different tools are required for study at different levels. To get useful answers we must ask the right questions.

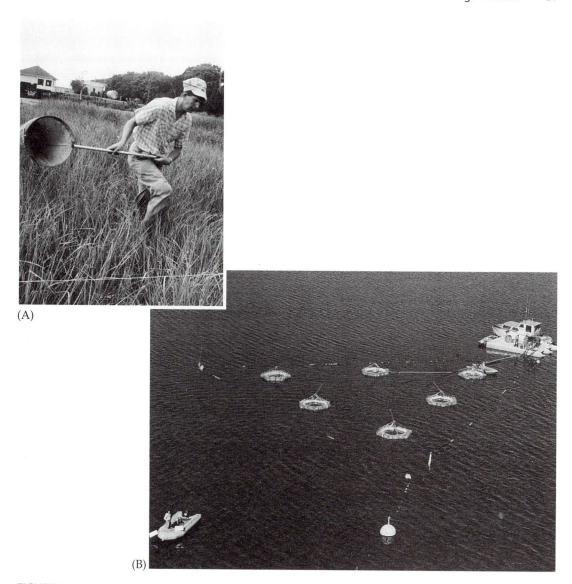

(A)

(B)

FIGURE 1
Different procedures and different tools are needed for the study of different levels of biological organization. (A) For a species-level study, such as sampling the kinds of insects in a salt marsh, a simple sweep net may be adequate. (Photograph by E. P. Odum.) (B) For a community-level study, such as determining the effects of a toxicant on community metabolism in a coastal bay, more elaborate and expensive equipment, like these floating mesocosms that enclose a whole water column, may be needed. (Photograph courtesy of David W. Menzel, Director, Skidaway Oceanographic Institute.)

Many times our attempts to solve an environmental problem fail, or even backfire, because the wrong question is asked, or the wrong level is focused on. For example, searching in the water for the cause of a fish kill may reveal the agent that killed the fish, but may not prevent future kills if the poison comes from some other part of the landscape.

In the next chapter, we will start our review of important ecological principles at the ecosystem level, which is a key middle level between you as an individual and the world in which you live.

About Models

How, then, do we begin to study something so complex and formidable as an ecological system? We begin just as we would begin the study of any complex situation—by describing simplified versions, which encompass only the more important or basic properties or functions. In science, simplified versions of the real world are called models. A **model** is a simplified formulation that mimics a real-world phenomenon so that complex situations can be comprehended and predictions made. In their simplest form, models may be verbal or graphic; that is, they may consist of concise statements or picture graphs. Although for the most part we shall restrict discussions in this book to informal models, it is worthwhile to consider the rationale of more formal models, because modeling is playing an increasingly important role in professional ecology and in science in general. Desktop computers and special modeling software now make it possible for someone with a minimal background in mathematics and physical science to model ecological situations.

In its formal version, a working model of an ecological situation would most likely have five components as follows (with certain technical terms used by modelers listed in parentheses):

1. **Properties** (P; state variables)
2. **Forces** (E; forcing functions), which are outside energy sources or causal forces that drive the system
3. **Flow pathways** (F), showing where energy or material transfers connect properties with each other and with forces
4. **Interactions** (I; interaction functions), where forces and properties interact to modify, amplify, or control flows
5. **Feedback loops** (L), where an output loops back to influence an "upstream" component or flow

Modeling usually begins with the construction of a diagram, or graphic model, which may take the form of a compartment diagram as illustrated in Figure 2. Shown are two properties, P_1 and P_2, which interact at I to produce or affect a third property, P_3, when the system is driven by forc-

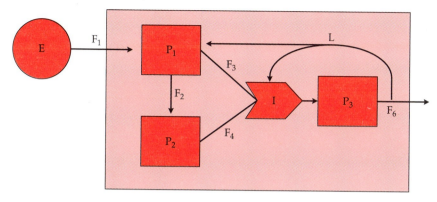

FIGURE 2

A systems diagram showing the five basic components that are of primary interest in modeling ecosystems: E, forces; P, properties, F, flow pathways; I, interactions; L, feedback loops.

ing function E. Five flow pathways are shown, with F_1 representing the input and F_6 the output for the system as a whole. Also shown is a feedback loop, L, signifying that a downstream output, or some part of it, is fed back, or recycled, to affect or control an upstream component or process.

Figure 2 could serve as a model for smog production in the air over Los Angeles (or any other large city). In this case, P_1 could represent hydrocarbons and P_2 nitrogen oxides, two components of automobile emissions. Under the driving force of sunlight, E, these interact to produce a new substance, photochemical smog. In this case the interaction function, I, is synergistic and augmentative in that P_3 is a more irritating pollutant than is P_1 or P_2 acting alone. A feedback loop would be appropriate for the model if it could be shown that as the concentration of smog in the air increases, the rate of production of new smog is either increased or decreased. Accordingly, feedback can be either positive or negative (and could be so indicated by a plus or minus sign on the diagram). In general, positive feedback speeds up a system or process (as a governmental subsidy might speed up economic development) while negative feedback slows down or maintains a process at a steady rate (as a zoning plan is supposed to control urban development in an orderly manner). Both types are common in nature, as we shall see.

Alternatively, Figure 2 could serve as a model for a grassland ecosystem in which P_1 represents green plants (e.g., the grasses) which convert sun energy, E, to food. P_2 could represent herbivorous animals that eat plants, and P_3, omnivorous animals that can eat either the herbivores or the plants. In this case the interaction function could represent several possibilities. It could be a "no-preference switch" if observation indicates that the omnivores feed on plants and animals according to availability. Or I could be specified to be a constant percentage value if the diet of the omnivores averages, say, 80 percent plant matter and 20 percent animal matter. Or I could be a seasonal switch if P_3 switches from plants to animals according to season.

These examples will suffice to show the tremendous versatility of model building, not only to provide simplified versions of the real world that help us understand it, but also to set up hypothetical test cases to answer "What if?" questions—for example, What would happen if this property were removed or another added, or that interaction changed, or that energy source reduced, or that feedback altered? To use and experiment with models for any theoretical or practical purpose, the chart models that we have been discussing must be converted to mathematical models by quantifying properties and drawing up equations for flows and their interactions. As we have already noted, computer software is now available that does the equation work, making it possible to deal with larger and more complex formulations—but this is a subject for advanced study. For now, we just need to understand how model builders go about their business (H. T. Odum 1971; Hall and Day 1977; Hannon and Ruth 1994).

Suggested Readings

*Allen, G. E., ed. 1980. Integrated pest management. Special issue of *BioScience* 30: 655–701.

*Allen, T. F. H., and T. B. Starr. 1982. *Hierarchy: Perspectives for Ecological Complexity*. University of Chicago Press, Chicago.

*Barrett, G. W., J. D. Peles, and E. P., Odum. 1997. Transcending process and the levels-of-organization concept. *BioScience* (in press).

*Belt, C. B., Jr. 1975. The 1973 flood and man's constriction of the Mississippi River. *Science* 189: 681–684.

Benson, A. J., and C. P. Boydstun. 1995. Invasion of the zebra mussel in the United States. In *Our Living Resources*, ed. T. LaRoe, pp. 445–448. National Biological Service, U.S. Department of the Interior, Washington, DC.

Bright, C. 1996. Understanding the threat of bioinvasions. Chapter 6 in *State of the World 1996*, ed. L. R. Brown, pp. 95–113. Worldwatch Institute, Washington, DC. (See also *Worldwatch* 8(4): 10–19, 1995.)

Fiebleman, J. K. 1954. Theory of integrated levels. *Brit. J. Phil. Sci.* 5: 59–66.

*Gardner, Gary. 1996. IPM (Integrated Pest Management) and the war on pests. *Worldwatch* 9(2): 21–27. (Chemical weapons are a losing proposition, so what are hopes for IPM?)

*Hall, C. A. S., and J. W. Day. 1977. Systems and models: Terms and basic principles. In *Ecosystem Modeling in Theory and Practice*, pp. 5–36. John Wiley, New York.

Hannon, M., and M. Ruth. 1994. *Dynamic Modeling*. Springer-Verlag, New York. (Graphical language software STELLA used to demonstrate developing and then running models of real-world systems.)

Higashi, M., and T. Ohgushi. 1990. Three interconnections of ecology. Chapter 14 in *Ecology for Tomorrow*. Special issue, *Physiol. Ecol. Japan* 27. (Hierarchical perspective, network perspective, and evolutionary perspective are the three interconnections.)

Hutchinson, G. E. 1964. The lacustrine microcosm reconsidered. *Am. Sci.* 52: 334–341. (Discusses the holological [wholes] and merological [parts] approaches as contrasting philosophies in the study of lakes and other complex systems.)

*Ludansky, M. L., D. McDonald, and D. MacNeill. 1993. Impact of the zebra mussel, a bivalve invader. *BioScience* 43: 533–544.

*Meentemeyer, V., and E. O. Box. 1989. Scale effects in landscape studies. In *Landscape Heterogeneity and Disturbance*, ed. M. G. Turner. Springer-Verlag, New York.

*Minnich, R. A. 1983. Fire mosaics in southern California and northern Baja California. *Science* 219: 1287–1294.

*Mooney, H. A., and J. A. Drake, eds. 1986. *Ecology of Biological Invasions in North America and Hawaii*. Ecological Studies. Springer-Verlag, New York.

*Murdoch, W. W., J. Chesson, and P. L. Chesson. 1985. Biological control in theory and practice. *Am. Nat.* 125: 344–366.

National Research Council. 1995. *Ecological Based Pest Management: New Solutions for a New Century*. National Research Council, Washington, DC.

Novikoff, A. B. 1945. The concept of integrative levels in biology. *Science* 101: 209–215.

Odum, E. P. 1977. The emergence of ecology as a new integrative discipline. *Science* 195: 1289–1293. (Ecology must combine holism with reductionism if applications are to benefit society.)

*Odum, H. T. 1971. The world system. Chapter 1 in *Environment, Power, and Society*, pp. 1–25. Wiley-Interscience, New York.

*O'Neill, R. V., D. L. DeAngelis, J. B. Waide, and T. F. H. Allen. 1986. *A Hierarchical Concept of Ecosystems*. Monographs in Population Biology No. 23. Princeton University Press, Princeton, NJ.

*Salt, G. W. 1979. A comment on the use of the term emergent properties. *Am. Nat.* 113: 145–148.

*Schlesinger, A. M. 1986. *The Cycles of American History*. Houghton Mifflin, Boston. (Picking up a theme from Henry Adams, Schlesinger discusses the apparent alternation of periods of conservatism and liberalism.)

Simon, H. A. 1973. The organization of complex systems. In *Hierarchy Theory*, ed. H. H. Pattee. George Braziller, New York.

*Stone, R. 1992. Researchers score victory over pesticides—and pests—in Asia. *Science* 256: 1272–1273. (When pesticide use actually increased rice pest outbreak by causing loss of natural enemies and development of resistant strains, a modified "integrated pest management" increased production and reduced pesticide use sharply.)

Turner, M. G. 1989. Landscape ecology: The effect of pattern and process. *Annu. Rev. Ecol. Syst.* 20: 171–197.

Urban, D. L., R. V. O'Neill, and H. H. Shugart. 1987. Landscape ecology. *BioScience* 37: 119–127.

*Indicates references cited in this chapter.

3

The Ecosystem

Sir Arthur Tansley (1871–1955) was an English botanist and one of the founders of the world's first ecological society, the British Ecological Society. His field of expertise was vegetation, but unlike many specialists, he had broad interests, including geology, psychology, and the philosophy of science and its methodology. He not only recognized that animals depend on plants, but also that plants depend on animals in many ways, and that both are closely knit together with the nonliving world. In 1935, he coined the term **ecosystem** for biotic and abiotic components considered as a whole. His selection of the word *system* clearly indicated that he was not thinking of *ecosystem* as a catchall word for everything that affects vegetation, but as a suitable name for an organized unit. The key concept, in his own words, "is the idea of progress towards equilibrium, which is never, perhaps, completely attained, but to which approximation is made whenever the factors at work are constant and stable for a long enough period of time" (Tansley 1935). Tansley's term did not come into general use in ecology until after his death, and only very recently has it become a part of everyday language.

The ecosystem level is the logical level around which to organize theory and practice in ecology because it is the lowest level in the ecological hierarchy (see Table 1 in Chapter 2) that is complete—that is, has all the components necessary for function and survival over the long term. For the same reason, we are hearing more about **ecosystem management** these days—we are moving up from trying to deal with the components separately, one at a time, to managing the system as a whole.

Ecosystem Models

As is the case for all kinds and levels of biosystems (biological systems), ecosystems are open systems—that is, things are constantly entering and leaving, even though the general appearance and basic functions may remain constant for long periods of time. Inputs and outputs (as first shown in the generalized system model of Figure 2 in Chapter 2) are an important part of the concept. As shown in Figure 1, a graphic model of an ecosystem can consist of a box that we can label the system, which represents the area we are interested in, and two large funnels that we can label **input environment** and **output environment**. The boundary for the system can be arbitrary (whatever is convenient or of interest), delineating an area such as a block of forest or a section of beach; or it can be natural, such as the shore of a lake where the whole lake is to be the system.

Energy is a necessary input. The sun is the ultimate energy source for the biosphere and directly supports most natural ecosystems within the biosphere. But there are other energy sources that may be important for many ecosystems, for example, wind, rain, water flow, or fuel (the major source for the modern city). Energy also flows out of the system in the form of heat and in other transformed or processed forms such as organic

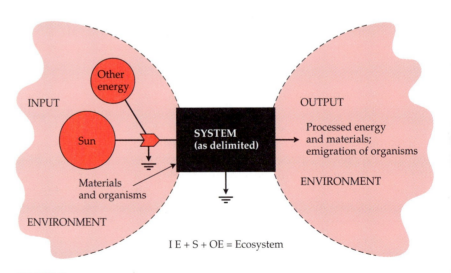

FIGURE 1
Model of an ecosystem as an open, thermodynamic nonequilibrium system, with emphasis on the external environment, which must be considered an integral part of the ecosystem concept.

matter (e.g., food and waste products) and pollutants. Water, air, and nutrients necessary for life, along with all kinds of other materials, constantly enter and leave the ecosystem. And, of course, organisms and their propagules (seeds and other reproductive stages) enter (immigrate) or leave (emigrate).

In Figure 1 the system part of the ecosystem is shown as a "black box," which is defined by modelers as a unit whose general role or function can be evaluated without specifying its internal contents. But we want to look inside this black box to see how it is organized internally and find out what happens to all those inputs. The contents, as it were, of an ecosystem are shown in model form in Figure 2 (and pictorially in Figures 4 and 5).

In Figure 2, the compartments (boxes) in the model have been given different shapes according to their basic functions, using the "energy language" symbols developed by H. T. Odum (1971), summarized in Figure 3. Circles represent renewable energy sources, bullet-shaped modules are autotrophs, hexagons are heterotrophs, tank-shaped boxes are storages,

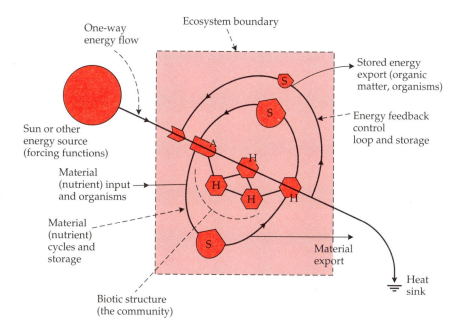

FIGURE 2
A functional diagram of an ecosystem, with emphasis on internal dynamics involving energy flow, material cycles, and storage (S), and food webs comprising autotrophs (A) and heterotrophs (H).

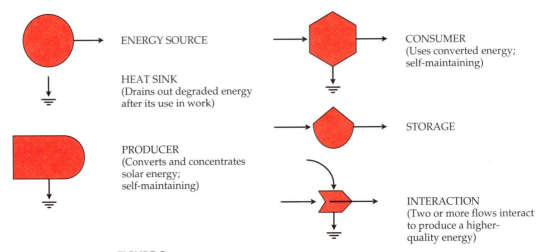

FIGURE 3
H. T. Odum's "energy language" symbols used in model diagrams in this book.

and arrows-into-ground are heat sinks (where heat is lost). This graphic language will be used in other models in this book.

Ecosystems have two major biotic components. First is an **autotrophic** (self-nourishing) component, able to fix light energy and manufacture food from simple inorganic substances (e.g., water, carbon dioxide, nitrates) by the process of photosynthesis. Generally, the green plants—vegetation on land, algae and water plants in aquatic habitats—constitute the autotrophic component. These organisms may be thought of as the **producers**. As shown in Figure 4, they form an upper "green belt" or stratum (layer) where the sun energy input is greatest.

The second major unit is the **heterotrophic** (other-nourishing) component, which utilizes, rearranges, and decomposes the complex materials synthesized by the autotrophs. Fungi, nonphotosynthetic bacteria and other microorganisms, and animals, including humans, constitute the heterotrophs, which concentrate their activities in and around the "brown belt" of soil and sediment below the green canopy. These organisms may be thought of as the **consumers**, since they are unable to produce their own food and must obtain it by consuming other organisms. In the model diagram of Figure 2, the autotrophic (A) and heterotrophic (H) components are shown linked together in a network of energy transfers called a **food web**. We will consider food webs in more detail in Chapter 4.

It is often useful to further subdivide heterotrophs according to the source of their food energy. Thus, we have the **herbivores**, which feed on

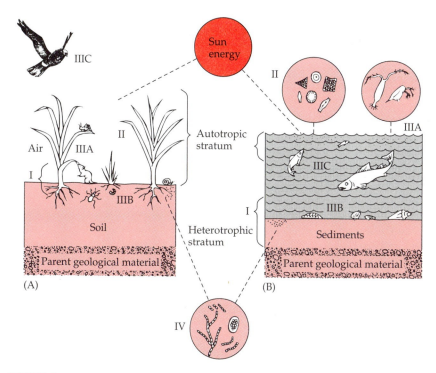

FIGURE 4

Sun-powered autotrophic ecosystems: comparison of the gross structure of a terrestrial grassland and an open-water ecosystem. Necessary units for function are: I. Abiotic substances (basic inorganic and organic compounds); II. Producers (vegetation on land, phytoplankton in water); III. Macroconsumers or animals: (A) direct or grazing herbivores (grasshoppers, meadow mice, etc., on land; zooplankton in water); (B) indirect or detritus-feeding consumers or saprovores (soil invertebrates on land, bottom invertebrates in water); (C) the "top carnivores" (hawks and large fish); IV. Decomposers (bacteria and fungi).

plants; the **carnivores** or **predators**, which feed on other animals; the **omnivores**, which feed on both plants and animals; and the **saprovores** (primarily microorganisms and fungi), which feed on decaying organic materials.

You will note that these ecological classifications of biotic components are based on modes of nutrition, that is, the principal source of energy utilized. Such ecological classifications should not be confused with taxonomic classifications of species (although there are parallels, since the three modes of nutrition—photosynthesis, ingestion, and absorption—predominate in the taxonomic kingdoms of plants, animals, and fungi respectively). Ecological classification is *one of function, not species*. Many species utilize more than one energy source, and still others are able to shift their mode of nutrition. For example, some kinds of algae are able to function either as autotrophs or as heterotrophs according to the availability of sunlight and organic matter.

The Meadow and the Pond

Terrestrial and aquatic ecosystems are contrasting types, and Figure 4 emphasizes their basic similarities and differences. Land ecosystems and water ecosystems typically are populated by different kinds of organisms (although some organisms, such as ducks and frogs, live in both ecosystems at different times or during different life-history stages). Despite wide differences in species composition, the same basic ecological components are present and function in the same manner in both ecosystems. On land, the predominant autotrophs are usually rooted plants, ranging in size from the grasses and other herbs that occupy dry or recently denuded sites to the large forest trees adapted to moist lands. Near shore in a pond or in other shallow water situations (wetlands, for example), rooted aquatic plants (e.g., cattails, water lilies, buttonbush) occur, but in the vast open-water stretches of ponds, lakes, and oceans, the autotrophs are microscopic suspended plants called **phytoplankton** (*phyto*, plant; *plankton*, floating), which include various kinds of algae, green bacteria, and green protozoa.

Because of size differences in plants, the **biomass** (living weight) or **standing crop** of terrestrial systems may be very different from that of aquatic systems. Plant biomass may be 10,000 or more grams of dry matter per square meter in a forest, in contrast to 5 grams or less in the open water of ponds, lakes, and oceans. Despite this biomass discrepancy, 5 grams of phytoplankton are capable of manufacturing as much food in a given amount of time as are 10,000 grams of large plants, given the same input of light energy and nutrients. This is because the rate of metabolism of small organisms is much greater per unit of weight than that of large organisms. Furthermore, large land plants such as trees are composed mostly of woody tissues that are relatively photosynthetically inactive; only the leaves photosynthesize, and in a forest, leaves comprise only 1 to 5 percent of the total plant biomass. Accordingly, the amount of living material (biomass) that you see on the landscape is not necessarily indicative of the rate of production of living material.

This is a good place to introduce the concept of **turnover** as a first step in relating structure to function. We can think of turnover as the ratio of the standing stock (that is, the amount present at any one time) of biotic

Fishing in an Empty Sea

It is extremely important that we take into account the replacement times for renewable resources, living or otherwise. We cannot continue for long to remove fish from the sea or water from a well faster than they are replaced, yet this is just what we are doing in many places. Theoretically, our economic system should correct this dead-end process, since as a resource becomes scarce, it should rise in price, which should cause people to consume less. In practice, however, a rise in price often increases the profits for those who fish or drill for water, so exploitation continues to the dead end (as, for example, nearly happened with the harvest of some species of whales), unless we intervene in the marketplace with political and legal action. There is nothing un-American or undemocratic in such intervention; we do it all the time when the public welfare, public interest, or environmental quality is threatened. As we continue to emphasize in this book, market economics works well for allocation of human-made goods and services, but not for many natural resources.

or abiotic components to the rate of replacement of the standing stock. For example, if the biomass of a forest is 20,000 grams per square meter (g/m^2) and the annual growth increment is 1000 grams, then the ratio 20:1 can be expressed as a **turnover time** or **replacement time** of 20 years. The reciprocal, that is, $1/20 = 0.05$, is the **turnover rate**. In a pond, the turnover time for phytoplankton would be measured in days rather than years.

The difference between land and water ecosystems in biomass and turnover time is reflected in the ways we obtain our food and fiber from them. On land, plant biomass tends to accumulate over time—a growing season for crops, many years for a forest—so it can be conveniently harvested when a large or perhaps a maximum standing crop has accumulated. Therefore, basic human food produced on land is in plant form (grains, vegetables, etc.). In contrast, the turnover at the autotrophic level in the sea is so fast that very little biomass accumulates. What does accumulate in the sea is animal biomass (fish, crabs, whales, etc.), so practically all the seafood that we harvest or grow in aquaculture farms is in animal form. An exception is the large seaweeds that are cultivated in Asia and Africa.

Heterotrophic Ecosystems

In natural and seminatural landscapes that contain a variety of ecosystems (e.g., forests, grasslands, farmland, lakes, ponds, streams), autotrophic and heterotrophic activity taken as a whole tends to balance; the organic matter produced is utilized in growth and maintenance over the annual cycle. Sometimes production exceeds use, in which case organic matter may be stored (as peat in a marsh, for example) or exported to another ecosystem or landscape (as in agriculture). In contrast, cities (and industrialized landscapes in general) consume much more food and organic matter than they produce, and are accordingly heterotrophic ecosystems. Figure 5 compares an oyster reef, one of nature's heterotrophic ecosystems, with a city; both must get their food and other energy from outside. Note that on a unit of area basis the city requires much more energy per day than does the oyster reef (70 times more in the example shown). There is nothing wrong or bad about our cities being heterotrophic—so long as they are linked with adequate autotrophic systems that can supply the food and other energy (not to mention raw materials) required and can also assimilate the large output of wastes produced by the city. The absolute dependence of the city on its natural and domesticated countryside and the concept of the city as a parasite were emphasized in Chapter 2.

This brings us back to the basic theme of this book: the natural environment as the life-support module for Spaceship Earth. Since, as already noted, nature's capacity to support our ever-expanding and demanding cities is being stretched to the limit in many places, it is time to think about redesigning cities to reduce the drain (the input management concept as outlined in Chapter 2). Recycling water and wastes, growing food on rooftops, and using solar energy directly to heat buildings and produce electricity are some of the things that need to be done on a larger scale than they are at present.

Opportunity Knocks

There are many ways in which market economics can help in the redesign of cities. There is money to be made in waste management (including reduction of toxic waste production), recycling, solar electricity, sustainable agriculture, development of hydrogen fuels, genetic engineering for pest and disease control, insulation of houses, environmental restoration—the opportunities are endless!

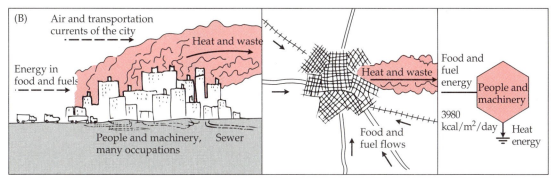

FIGURE 5
Heterotrophic ecosystems. (A) One of nature's "cities," an oyster reef is
dependent on the inflow of food energy from a large area of surrounding
environment. (B) An industrialized city is maintained by a huge inflow
of fuel and food, with a correspondingly large outflow of waste and heat.
Its energy requirement, on a square meter basis, is about 70 times that of
the reef: about 4000 kcal per day, or about 1.5 million kcal per year. (After
H. T. Odum 1971.)

Abiotic Components

Shown (in a simplified manner) in Figure 2 are the two basic abiotic functions that make the ecosystem operational, namely, **energy flow** and
material cycles. Energy flows from the sun or another external source,
through the biotic community and its food web, and out of the ecosystem
as the heat, organic matter, and organisms produced in the system.
Although energy may be stored and utilized later, energy flow is one-way
in the sense that once energy has been utilized—that is, converted from
one form to another (sunlight to food, for example)—it cannot be used

again; sunlight must continue to flow in if food production is to continue. Why this is so will be explained in Chapter 4. In contrast, chemical materials—elements and compounds—can be used over and over without loss of utility. In a well-ordered ecosystem, many of these materials cycle back and forth between abiotic and biotic components. These **biogeochemical cycles** are the subject of Chapter 5.

Of the large number of elements and simple inorganic compounds found at or near the surface of the earth, a certain few are essential for life. These are called **biogenic substances** or **nutrients**. As you might expect, these substances tend to be retained and recycled within living systems to a greater extent than nonessential ones. Carbon, hydrogen, nitrogen, phosphorus, and calcium, among others, are required in relatively large amounts and hence are designated as **macronutrients**; these occur abundantly in simple compounds, such as carbon dioxide, water, and nitrates, that are readily available to organisms. They also occur in chemical forms that are not readily available. Nitrogen in gaseous form in the air, for example, is not available to plants until it is converted to inorganic salt form (nitrate, ammonia) by specialized microorganisms or by other means (see Chapter 5). Phosphorus in the soil may occur in chemical forms unavailable to roots of plants; whether your tomato plant will have enough phosphorus depends not on the total amount in your soil but on the amount in available form. When you have your garden soil tested, the laboratory will report to you the amount of available nutrients and tell you how much you may need to add to ensure adequate amounts for your vegetables.

Other elements, no less vital than the macronutrients but required only in small amounts by organisms, are known as **micronutrients** or **trace elements**. There are a dozen or so of these that are essential to plants and most animals; they include a number of metal ions such as iron, magnesium, manganese, zinc, cobalt, and molybdenum. Still others are known or suspected to be essential for particular groups of organisms. The kinds and amounts of trace minerals needed for human health are not completely known, and this is currently a subject of much research and controversy. While 50–150 kilograms of each macronutrient such as nitrogen, phosphorus, and potassium is required to produce a good crop of corn on a hectare of land, less than 0.1 kilogram of most micronutrients is needed. However, because many micronutrients are scarce at the earth's surface, shortages do occur and can hamper the productivity of an ecosystem just as much as would a shortage of macronutrients. Without molybdenum, for example, the microorganisms mentioned above are unable to transform atmospheric nitrogen into ammonia and nitrates usable by plants.

The carbohydrates (e.g., sugars, starches, and cellulose), the proteins (including amino acids) and the lipids (e.g., fats and oils), which make up the bodies of living organisms, are also dispersed widely in nonliving forms in the environment. These and hundreds of other complex compounds make up the **organic component** of the abiotic environment. The important roles they play as feedback regulators will be discussed later.

As the bodies of organisms decay they become dispersed into fragments and dissolving materials, collectively called **organic detritus** (product of disintegration, from the Latin *deterere*, to wear away; in geology the word *detritus* also refers to products of rock disintegration). Since the biomass of plants is usually greater than that of animals, and since plants usually decay more slowly than animal remains, detritus of plant origin is usually more prominent than that of animal origin in the terrestrial environment. Organic detritus not only is a food source for saprovores both on land and in water, it also improves soil and sediment texture and enhances the retention of water and minerals, which is why mulch and compost are good for your garden.

Ecologists often use the shorthand notations **DOM** (dissolved organic matter) and **POM** (particulate organic matter) for the two forms of detritus. In the ocean DOM is a major reservoir of organic carbon and is important in the global carbon cycle (Benner et al. 1992). It is a food source for bacteria, and also affects light penetration. Under some conditions DOM is converted into POM, which can be consumed by filter-feeding animals.

As the breakdown of organic matter proceeds, materials called **humus** or **humic substances** are formed that are often resistant to further decay, which means that they may remain for some time as a structural part of the ecosystem. Humus is the dark yellow-brown amorphous or colloidal substance readily visible in soils and sediments and suspended in the water of streams and lakes (especially noticeable in swamp or bog water). Humic substances are difficult to characterize chemically. For those of you who have had a course in organic chemistry, we can say that they consist of chains of aromatic or phenolic benzene rings with side chains of nitrogen complexes and carbohydrate residues. The role that humic substances play in ecosystems is not fully understood, but we do know that they can either stimulate or inhibit plant growth, depending on other environmental conditions. Under certain conditions, such as existed in past geological ages, organic matter becomes fossilized, first as peat and then as coal, petroleum, and other fossil fuels on which our industrial societies now depend. Unfortunately, the by-products of industry, including these petrochemicals (made from petroleum), have become increasing voluminous and increasingly poisonous in recent decades, and the technology of waste management and waste reduction has lagged far behind our ability to produce toxic substances. The problem of toxic wastes will be discussed in Chapter 5.

We now come to the third category of the abiotic part of the input environment of ecosystems—the physical factors that determine the conditions of existence for the biotic community. Climate (e.g., temperature, rainfall, and humidity), the physicochemical nature of soil and water (e.g., salinity and pH), and the underlying geological substrata are some of the major features that determine the kinds of organisms present and, indirectly, how they are organized into communities and how well they are able to utilize available energy and resources.

Gradients and Ecotones

The biosphere is characterized by a series of gradients or zonations of physical factors. Examples are temperature gradients from the Arctic to the tropics and mountaintop to valley; moisture gradients from wet to dry along major weather systems; and depth gradients from shore to bottom in bodies of water. Frequently, conditions and the organisms adapted to them change gradually along a gradient, but often there are points of abrupt change, known as **ecotones**, as, for example, prairie–forest junctions or intertidal zones on a seacoast.

An ecotone is not simply a boundary or an edge; the concept assumes the existence of active interaction between two or more ecosystems (or patches of ecosystems), which results in the ecotone having properties that do not exist in either of the adjacent ecosystems (Decamps and Naiman 1990). In addition to external processes causing discontinuities in gradients, internal processes such as sediment traps, root mats, special soil–water conditions, inhibitory chemicals, or animal activity (dam building by beavers, for example) may maintain an ecotone as distinct from bordering communities (W. E. Odum 1990).

Sometimes ecotones are populated by more kinds and larger numbers of birds and game animals than can be found in the interior of the adjoining, more homogenous communities. Wildlife managers speak of this as the **edge effect** and often recommend special plantings between field and forest, for example, to increase the numbers of these animals. However, a sharp edge, such as that between a clear-cut and an uncut forest, may be poor habitat, and the large amount of edge in a fragmented, domesticated landscape usually reduces diversity. As we will see below, humans tend to chop up the landscape into blocks and strips with sharp edges, thus more or less doing away with natural gradients and ecotones.

The Biotic Community: Habitats and Niches

We are all aware that the kinds of organisms to be found in both rural and urban areas in particular parts of the world depend not only on the conditions of existence—that is, hot or cold, wet or dry—but also on geography. Each major land mass and each major ocean has its own special flora and fauna. Thus, we expect to see kangaroos in Australia but not elsewhere, or hummingbirds and cacti in the New World but not the Old World. And the different continents are the original homes of different races of humans and their domesticated plants and animals. From the standpoint of the overall structure and function of ecosystems, it is important only that we realize that the biotic units (species, etc.) available for incorporation into communities varies with the region (note that in Table 1 in Chapter 2, "Biogeographic region" is listed as a major level in the ecological hierarchy).

What is not so well understood is that ecologically similar species, or **ecological equivalents**, are found in different parts of the globe where the physical environment is similar. The grass communities in the temperate, semiarid part of Australia are composed of different species than those in a similar climatic region of North America, but they perform the same basic function as producers in the ecosystem. Likewise, the grazing kangaroos of the Australian grasslands are ecological equivalents of the bison and antelope (or the cattle that replaced them) on North American grasslands, since they have a similar functional position in the ecosystem. Ecologists use the term **habitat** to mean the place where a species can be found, and the term **ecological niche** to mean the ecological role of an organism in its community. The habitat is the organism's "address," so to speak (where it lives), and the niche is its "profession" (how it lives, including how it interacts with and is constrained by other species). Thus we can say that the kangaroo, bison, and cow, although not closely related genetically, occupy similar niches (i.e., are ecological equivalents) when present in grassland ecosystems. (How to delimit and measure width, overlap, and other niche dimensions is a subject much discussed by ecologists.)

As already mentioned, humans greatly alter the composition of biotic communities wherever they settle, not only by modifying the environment but by removing species and introducing new ones, both inadvertently and on purpose. Whether an introduction involves the replacement of one species by another in the same niche or the filling of an unoccupied niche, the overall effect on the functioning of the ecosystem may be neutral, beneficial, or detrimental. When Midwestern prairies were converted to agricultural fields (as European settlers replaced the native Indians) the native prairie chicken was unable to adapt to the greatly altered environment, but the introduced ring-necked pheasant, which was adapted to the farming country of Europe, thrived in the altered (domesticated) landscape. As far as the hunter is concerned, the "game bird niche" has been adequately filled, since shooting pheasants is as good if not better sport than shooting prairie chickens. Too often, however, introduced species become pests, as discussed in Chapter 2. This is why customs agents are so fussy about what kinds of plants or animals you might be trying to bring from one country to another, and especially from one continent to another.

Especially grave problems often result when domesticated plants (**cultivars**) and animals escape back to nature and become pests because of the absence of both human and natural controls. For example, on some of the Hawaiian Islands, **feral** (once domestic but readapted to the wild state) goats have had a more severe impact on soil, flora, and fauna than have bulldozers. Some of our most persistent "weeds" are escapees that have run amuck.

The Biotic Community: Species Structure

Nature, just like well-ordered human societies, has its specialists and its generalists when it comes to niches or professions. There are insects, for example, that feed only on a special part of one species of plant, while other insects may be able to feed on dozens of different species of plants. In general, specialists are efficient in the use of their resources (since all of their adaptations and behaviors are concentrated on a specialized way of life); therefore, they often become abundant when their resources are in ample supply. But the specialist is vulnerable to changes or perturbations that adversely affect its narrow niche. Since the niches of nonspecialized species tend to be broader, they are more adaptable to changing or fluctuating environments, even though they are never as locally abundant as the specialists.

Most natural communities contain so many species and varieties (including both specialists and generalists) that it would be impossible to catalog all the kinds of plants, animals, and microbes to be found in any large area such as a lake or tract of forest. Fortunately, it is not necessary to know all the species to assess the roles of species in community structure and function, because a characteristic and consistent feature of natural communities is that they contain *a comparatively few species that are common* (represented by large numbers of individuals or a large biomass) and *a comparatively large number of species that are rare* at any given place and time. A tract of hardwood forest, for example, may contain 50 or more species of trees, of which half a dozen or fewer account for 90 percent of the timber. Accordingly, we can focus our attention on the few common species, knowing that they will account for most of the action.

A tabulation made by an ecology class in a prairie ecosystem illustrates this general picture. As shown in Table 1, two species comprised 36 percent, nine species 84 percent, and the remaining twenty species only 16 percent (less than 1 percent each) of the total stand of vegetation. The few common species in a particular community grouping are called **ecological dominants**.

Although dominants may account for most of the standing crop and community metabolism, this does not mean that the rare species are unimportant. Species that exert some kind of controlling influence, whether or not they are dominants, are called **keystone species**. In the aggregate, rare species have an appreciable impact, and they determine the diversity of the community as a whole. Should conditions become unfavorable for the dominants, rarer species adapted to or tolerant of the new conditions may increase in abundance and take over vital functions. For example, when a blight (fungal disease) killed the dominant chestnut trees in the southern Appalachians in the 1940s, several species of oaks

Table 1 Species Structure of the Vegetation of an Ungrazed Tall-Grass Prairie in Oklahoma	
Species	**Percent of Stand[a]**
Sorghastrum natans (Indian grass)	24
Panicum virgatum (Switch grass)	12
Andropogon gerardi (Big bluestem)	9
Ailphium laciniatum (Compass plant)	9
Desmanthus illinoensis (Prickleweed)	6
Bouteloua curtipendula (Side-oats grama)	6
Andropogon scoparius (Little bluestem)	6
Helianthus maximiliana (Wild sunflower)	6
Schrankia nuttallii (Sensitive plant)	6
20 additional species (average 0.8% each)	16
Total	100

Source: Rice 1952, based on 40 one-square-meter quadrat samples taken by an ecology class

[a]In terms of percent cover of total of a 34% area coverage of soil surface by the vegetation. Figures are rounded off to nearest whole number.

gradually replaced the chestnuts, and timber density returned to preb-light levels in about 50 years.

The "Weed" Problem

By promoting the abundance of species with high economic value and utility for human use, agricultural, lawn care, and forestry practices greatly reduce species diversity in intensively cultivated or managed areas, often to the point of creating a situation that is vulnerable to any change, as already noted. The species structure of a cultivated grain field in midseason is shown in Table 2. In this case no herbicides or mechanical weeding were applied, so about 7 percent of the plant community consists of ten other species that have managed to invade the millet crop. A great deal of energy and expensive chemicals would be required to eliminate all the "weeds" and maintain a true **monoculture** (a one-species crop, lawn, or forest). Pests tend to zero in on monocultures and often develop resistance to herbicides and pesticides, so that an even more toxic chemical is required to do the job. Furthermore, frequent plowing and heavy applications of chemicals can cause serious soil erosion and water pollution.

Many scientists now question whether increasing the intensity of farming in an effort to get a little more yield does more harm than good. Some studies are showing that the presence of weeds in moderation may be ben-

Table 2 Species Structure of the Vegetation of a Cultivated Millet Field in Georgia	
Species	**Percent of Stand**[a]
Panicum ramosum (Brown-topped millet)	93
Cyperus sp. (Nut sedge)	5
Amaranthus hybridus (Pigweed)	1
Digitaria sanguinalis (Crabgrass)	0.5
Cassia fasciculata (Sicklepod)	0.2
6 additional species (average 0.05% each)	0.3
Total	100.0

Source: Barrett 1968

[a]In terms of percent dry weight of above-ground plants based on 20 quarter-square-meter quadrat samples taken in late July.

eficial to a crop by providing a habitat for useful insects or by improving soil conditions. Other studies are showing that mixtures of crops (polycultures) may produce more food or other products per unit of area than monocultures do. Agroecologists have recently become interested in taking a new look at traditional agricultural practices such as the ancient Indian corn–bean–squash crop mixtures that are still in use in Mexico and Central America (Figure 6). All of these interesting possibilities are now being researched at universities and experimental stations.

We should think of a "weed" or even a "pest" not so much as an undesirable species that should be wiped off the face of the earth, but rather as a species that is in the wrong place at the wrong time. A plant that is choking the flowers or vegetables in your garden may turn out to be a very useful member of a fallow field plant community, or an attractive wildflower along the roadside (crabgrass, dandelions, and morning glories are examples).

FIGURE 6
Modern contour plowing methods are applied to traditional Indian polyculture in this garden of corn (maize) and beans growing in the highlands of Guatemala. Such traditional agriculture may be better suited to the needs of many developing nations than the expensive, energy-intensive monocultures of the industrialized nations. (Photograph by Robert E. Ford/Terraphotographics.)

Landscape Ecology and the Human Domain

It is interesting that where humans create what might be called a "production ecosystem" that maximizes food or fiber yield, we find monocultures convenient to manage, especially with machinery. On the other hand, when we create "protective ecosystems" around our homes, we tend to go in for increased diversity. In a study of the vegetation of residential districts of Madison, Wisconsin (G. J. Lawson et al., unpublished data), 150 species of trees and shrubs (many of them exotic introductions) were identified, as compared with only about 30 species in a nearby forest preserve. The diversity of grasses, flowers, and small songbirds was also much greater in the city suburbs than in the natural forest. And it was found that the average suburbanite puts as much fertilizer and labor (per unit of area) into caring for his lawn as a farmer does in producing a crop of corn!

Because humans tend to create a **fragmented** or **patchy landscape** that is a **mosaic** of many ecosystem types, research and land use planning in human-dominated areas are best organized around the landscape and regional levels of the ecological hierarchy (see Table 1 in Chapter 2). A pioneer of landscape ecology as we know it today was Alexander von Humboldt (1769–1859), an explorer–naturalist–geographer and one of Charles Darwin's mentors. In his explorations of South America, Humboldt was impressed with the interdependence of human cultures (both native and European) and local environments. Philosophically, he was a holistic ecologist well before the word *oecology* was coined by Ernest Haeckel in 1886. For example, he wrote about the "chain of connections" in which all elements of earthly life, including humans, are mutually dependent. (For more on Humboldt, see Jordan 1981 and Sachs 1995.)

When aerial photos became generally available in the 1930s and 1940s, landscape-level study blossomed, first in Europe, where the landscape is not only fragmented but also relatively stable in terms of its patchwork patterns as compared with the rapidly changing patterns in the Americas. In recent years the development of computerized Geographic Information Systems (GIS) technology has greatly increased our ability to describe, compare, and model landscape mosaics as a basis for improving management and planning. (For more on this developing field, see reviews by Turner 1989 and Forman 1995b, and books by Naveh and Lieberman 1984, Zonneveld and Forman 1990, and Forman 1995a.)

In a patchy landscape, **patch size** and **patch shape** are important factors in determining what species of animals are able to survive. The smaller a patch, the greater the negative effects of fragmentation and the less positive the influences of edge. Figure 7 shows how a forested region has been fragmented by agricultural and suburban development in central Maryland. Agricultural development (Figure 7A) tends to reduce the forest to scattered woodlots or other patches with sharp edges (landscape "sharpening" according to Janzen 1987), while suburban development (Figure 7B) reduces the forest to strips and corridors along roads and streams (landscape "shredding" according to Feinsinger 1994). In either case, most of the patches and strips are too small to support forest-interior songbirds such as wood thrushes and some warblers. So-called edge species, such as blackbirds, blue jays, house wrens, and house finches, thrive in this fragmented landscape (Whitcomb et al. 1981). In an interesting experimental study on the effect of fragmentation, Kruess and

(A)   (B)

FIGURE 7
Fragmentation of the landscape in a forested region of central Maryland in an agricultural area (A) and in an expanding urban area (B). Black areas represent forest; white areas, cropland or open fields; gray or cross-hatched areas, residential development.

Hot Spots in a Cool Matrix

The concept of "hot spots" was introduced in Chapter 1 in reference to cities as very energetic patches scattered about in larger but less energetic life-supporting environments. Not only are many nonhuman species so distributed, but whole communities of species aggregate in favorable locations, creating a highly patchy distribution. For example, national breeding bird censuses carried out over a period of years have revealed that 20% of the census routes contain more than 50% of the individual birds (Brown et al. 1995). Another example is soil, in which as much as 90% of the activity of soil organisms occurs in small aggregates and root zones that constitute less than 10% of the total soil volume (Coleman 1995).

Tscharntke (1994) found that small islands of red clover were colonized by herbivorous insects much faster than by parasitoid and predatory species as compared with large plots, thus reducing natural parasitic and predator control.

The antidote for fragmentation and too much edge is to set aside large preserves before development becomes intense, and to maintain corridors, such as riparian zones and floodplains along streams, so that animals can move freely from patch to patch (Harris 1984). Such "thinking ahead" land use planning will contribute to sustaining the aesthetic quality of life as well as life-support systems for suburbanites.

Variety Is More than Just the Spice of Life

Because **diversity** is both a fascinating and important subject, it merits detailed and systematic study. First, we need to recognize two components of diversity: (1) the **richness** or **variety** component, which can be expressed as the number of kinds (species, genetic varieties, land use categories, etc.) per unit of space or as a ratio of kinds to numbers, and (2) the **relative abundance** component, or the apportionment of individuals among the kinds. Thus, two communities could have the same number of species but be very different in terms of the relative abundance or dominance of each species. For example, two communities might each have ten species, but one community might have the same number of individuals in each species, while most of the individuals in the other community might belong to one dominant species.

A convenient way to express and compare diversity is to calculate diversity indices based on the ratio of parts to the whole, or n_i/N, where n_i is the number or other **importance value** (biomass, productivity, etc.) of each component (species, for example) and N is the total of importance values. The percent of stand or percent of total, as shown in Tables 1 and 2, become such ratios when the decimal is shifted two places to the left (24 percent becomes 0.24, for example). To compare the degree of dominance, the **Simpson index** is often used. It is calculated by squaring the ratios for each kind and summing them, thus: $D = \Sigma(n_i/N)^2$. Table 3 uses the Simpson index to compare variety and dominance in the vegetation of the prairie and the millet field in Tables 1 and 2. Another widely used index is the **Shannon index**, $H = -\Sigma n_i/N \log_e n_i/N$, which is an approximation of a function originally proposed as a general measure of information.

Another way to compare diversity in different communities is to plot **dominance–diversity** curves, as shown in Figure 8, in which the importance of each species (or other component) is plotted in sequential order from highest (most abundant) to lowest (least abundant). Shown are profiles for the trees of three forests. The high dominance–few species pattern of a subalpine forest contrasts with the low dominance–many species pattern of a tropical rain forest. This range just about covers what one finds in nature, whether one is assessing the diversity of an ecological group (e.g., producers or parasites) or a taxonomic grouping (e.g., birds, insects, or fishes). Diversity tends to be lowest where physical conditions are limiting to life (e.g., the Arctic, a salt lake, a polluted stream) and highest in benign environments where conditions are favorable for a large variety of life. What has come to be known as the "intermediate disturbance hypothesis" (Sousa 1984) holds that moderate disturbance by forces outside the community can increase diversity in any community regardless of its position on an environmental gradient.

Table 3	Comparison of Species Variety and Dominance in the Vegetation of a Prairie and a Cultivated Millet Field	
	Number of Species	Dominance (Simpson Index)
Natural prairie	29	0.13
Cultivated millet field	11	0.89

Based on data in Tables 1 and 2.

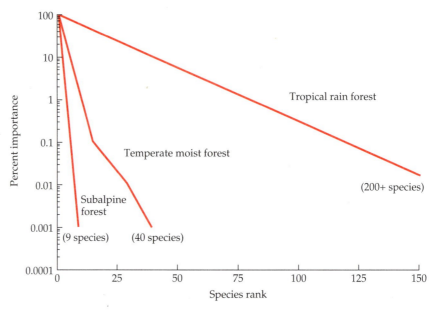

FIGURE 8
Dominance–diversity curves of three contrasting forests. Importance values for each species, ranked from the most to the least abundant, are based on net primary production for the two temperate forests and on aboveground biomass for the tropical rain forest. (After Hubbell 1979.)

There are two proverbs arising out of generations of human experience that are related to diversity and are often heard in everyday conversations. One is, "Variety is the spice of life," and the other, "Don't put all your eggs in one basket." Variety among living organisms certainly enriches our lives, but it also has a very practical value. It is much safer to have more than one kind of organism that can carry out a vital function. We never know when a rare species of plant or animal may provide a new drug or be needed to replace a common one that falls victim to disease.

Currently there is much concern not only about loss of **species diversity**, but also about loss of **genetic diversity** due to human activities. As the twentieth century comes to a close, concern about preservation of biotic diversity is reaching public and political levels (Wilson 1988). In the United States, special efforts are being made by the governmental, legal, and private sectors to identify and protect **endangered species** so as to maintain a high diversity of wild species. Similar efforts are being made

The Loss of Biodiversity

The word *biodiversity* has become almost a "buzzword" for our concern over the loss of species. As is made clear in this chapter, our concern should go beyond the species level to include the loss of functions and niches all up and down the hierarchical scale. In this context Wilcox (1984) defines **biodiversity** as "the variety of life forms, the ecological roles they play and the genetic diversity they contain."

It is especially important to maintain redundancy in an ecosystem—that is, to have more than one species or species group capable of carrying out major processes or providing important links in the food web. In assessing the effects of the removal or addition of species, it is important to know if they might be keystone species, a term introduced earlier in this chapter. Chapin et al. (1992) define a keystone species as "a functional group without redundancy." The loss of such a species or group of species will cause major changes in community structure and/or ecosystem function. And, as is so often the case, too much of something can also become detrimental. As discussed in Chapter 2, an exotic invader species added to a naturally diverse ecosystem often becomes a keystone species that reduces diversity.

In reviewing his long-term lake acidification experiments, Schindler (1990) found that there was a great deal of redundancy in the phytoplankton, so that primary production was not affected by moderate acidification of a lake. As acid sensitive species dropped out, they were replaced by others in the same functional niche. But there was much less redundancy among the species of the lake bottom (benthos). Food chain structure and the function of the whole lake were affected by a decline in certain crustacean species because no other species were present that could replace them as a link in the food chain between primary producers and fish. Accordingly, these crustaceans proved to be true keystone species.

to establish **gene banks** to preserve as many varieties of food plants as possible in case the ones in common use fail for whatever reason.

It has recently been noted that traditional (preindustrial) agriculture as practiced in many of the less developed countries in Latin America and

Asia provides the best of all gene banks because of the very wide variety of food crops and other species maintained in cultivation (Rhoades 1990). Traditional agriculture has many other advantages over modern methods in some circumstances, as we shall see in Chapter 5.

Landscape Diversity

An area can have low-diversity monocultures such as crops or tree plantations, yet still have high diversity in the landscape as a whole if there are a variety of more natural ecosystems scattered about. Figure 9 illustrates a case in which landscape diversity is high despite the fact that one species, pine, occupies 30 percent of the landscape. As shown, landscape diversity would drop drastically if 50 or 75 percent of the area was converted to pine plantations.

It now seems reasonably certain that **redundancy** (repetition) and **resilience stability** (the ability to recover rapidly from disturbance) are both enhanced by the presence of many different species in the landscape. Diversity, therefore, is well worth preserving wherever possible, since we are going to need all the help we can get from nature to promote the recovery of the many landscapes we have abused. Whether a high species diversity increases **resistance stability**—the ability of the ecosystem to remain the same (stable) in the face of disturbance—is a question much debated by ecologists. Recent field experiments by Tilman and associates (1996) have indicated that a high diversity of species in grassland communities does indeed increase stability during drought. This question will be discussed further in Chapter 7.

Kinds of Ecosystems

The human mind seems to require, or we might even say delights in, orderly categorization or classification when it comes to dealing with a large variety of entities—books in a library, or species, or jobs in the office or factory. Ecologists have not agreed on any one classification scheme for ecosystem types, or even upon what would be a proper basis for one—and this is as it should be, since many different arrangements can be instructive. Either structural or functional characteristics can serve as a basis for classifications. Let us cite three useful arrangements.

The biome classification scheme (refer back to Chapter 2 for an explanation of the term *biome*) is a widely used system based on conspicuous, ever-present structural macrofeatures. On land, vegetation usually provides an easily recognizable macrofeature that "integrates," as it were,

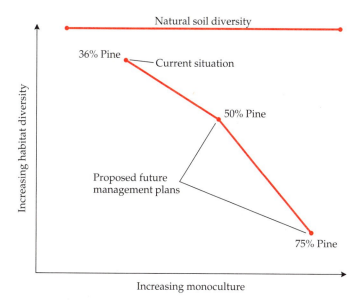

FIGURE 9
Landscape diversity in the 300-square-mile National Environmental Research Park at the Savannah River site, where one-third of the area is now in monoculture pine plantations. Because the plantations are intermixed with a wide variety of more diverse habitats, landscape diversity is high and approaches natural soil diversity. Landscape diversity would decrease greatly if pine plantations were extended to 50 or 75 percent of the area. (After Odum 1982; data from Odum and Kroodsma 1977.)

organisms, soils, and climate. In aquatic environments, where plants may be inconspicuous, dominant physical features provide the basis for recognition and classification. There are also the special and important human-altered or "domesticated" ecosystems such as cities and agricultural fields. The major ecosystem types of the biosphere are reviewed and illustrated in Chapter 8. Another widely used classification scheme uses **ecoregions**, based on a combination of natural and geographic attributes, as, for example, "the Appalachian Ecoregion" or "the Mississippi Delta Ecoregion" (see Bailey, 1995).

Energy provides an excellent basis for a functional classification of ecosystem types since it is the major common denominator for all ecosystems, natural, human-altered, or human-made. An energy-based classification will be presented in Chapter 4, after the basic laws governing energy behavior have been outlined.

The Gaia Hypothesis

James Lovelock is one of those rare but important persons who are not satisfied with achieving success and financial independence in one profession, but who aspire to higher intellectual levels. Trained as a physical scientist, he developed an analytical technique called the *electron capture detector* that greatly increased the chemist's ability to detect extremely small (trace) amounts of chemical substances. This technology led to the discovery that pesticides and other highly toxic residues were present in creatures all over the earth, from penguins in the Antarctic to mother's milk in the United States. Such revelations inspired Rachel Carson's influential book *Silent Spring* (1962) by providing the evidence needed to justify her concern about the long-term adverse consequences of the ubiquitous presence of human-made toxic chemicals. In the mid-1960s, after a period with NASA, Lovelock retired from the industrial world to a cottage in rural England to embark upon a second career. He devoted himself to testing in a logical and scientific way the ancient concept of Gaia, the Greek goddess "Mother Earth," and the general theory that organisms do not just passively adapt to physical conditions but actively interact to modify and control the chemical and physical conditions of the biosphere. For the next decade or so, he became a student of astronomy, cosmology, biology, and other disciplines that would enable him, in his own words, "to pursue an interdisciplinary journey in my quest for Gaia."

One of Lovelock's teachers and colleagues was Lynn Margulis, a prominent contributor to ideas about the origin of life. Together they published a series of articles summarizing the evidence for biological control of the physical environment and the important role of microorganisms in this control. The late Alfred Redfield (who worked at the Woods Hole Oceanographic Institution for many years), another interdisciplinary thinker with a remarkable breadth of expertise in both biological and physical science, also contributed independently to the concept (Redfield 1958). In 1979, Lovelock published a readable little book entitled *Gaia: A New Look at Life on Earth*, which, in his own words, "is a personal account of a journey through space and time in search of evidence to substantiate this model of earth."

The **Gaia hypothesis**, in Lovelock's words, states that "the biosphere is a self-regulating entity with the capacity to keep our planet healthy by controlling the chemical and physical environment." In other words, the earth is a superecosystem (but not a superorganism, since its development is not genetically controlled) with numerous interacting functions and feedback loops (recall the general model in Figure 2 in Chapter 2) that moderate extremes of temperature and keep the chemical composition of the atmosphere and the oceans relatively constant. Also—and this is the most controversial part of the hypothesis—the biotic community plays the major role in biospheric homeostasis, and organisms began to establish control soon after the first life appeared more than three billion years ago. The contrary hypothesis, of course, is that purely geological (abiotic) processes produced conditions favorable for life, which then merely adapted to these conditions.

The question, then, is: Did physical conditions evolve (change) first, and then life, or did both evolve together? The first, or primary, atmosphere, most scientists agree, was formed from gases rising from the hot core of the earth (e.g., through volcanoes), a process called *outgassing* by geologists. While volcanoes continue to affect climates (although on a lesser scale), the secondary atmosphere, the one we have now, is a biological product, according to the Gaia hypothesis. This reconstruction, as it were, began with the first life, the primitive microbes that do not require oxygen—the **anaerobes**. When the green anaerobic microbes began to put oxygen into the air, the plants and animals that require gaseous oxygen, the **aerobes**, evolved and the anaerobes retreated to the oxygenless depths of soils and sediments, where they continue to thrive and play a major role in various ecosystems. Perhaps as important was the cooling of the earth that resulted from the removal of carbon dioxide from the atmosphere by limestone-forming marine organisms.

Comparison of the atmosphere of the earth with that of the planets Venus and Mars, where if there is life it is restricted to anaerobic microbes, provides strong indirect evidence for the Gaia hypothesis. As shown in Figure 10, the low carbon dioxide–high oxygen and nitrogen atmosphere of the earth is completely opposite from the conditions on nearby planets. Since photosynthesis, which evolved soon after the first appearance of life, removes carbon dioxide from and adds oxygen to the atmosphere, and since this autotrophic activity in the past often exceeded the reverse gaseous exchange of respiration (witness fossil fuel deposition), and in view of permanent carbon dioxide removal by limestone formers, it is logical to conclude that the biotic community is responsible for the buildup of oxygen and reduction in carbon dioxide over time. Until recently, many geochemists assumed without much direct evidence that our oxygen came solely from the breakdown of water vapor and the escape of hydrogen into space, leaving an excess of oxygen behind. It is also difficult to explain how gaseous nitrogen could accumulate in the atmosphere in the absence of life. Without contrary biological transformations, nitrogen would go to its most stable form, nitrate ion dissolved in the oceans.

The nitrogen cycle, described in Chapter 5, clearly demonstrates that the biotic community does not just borrow gases from the atmosphere

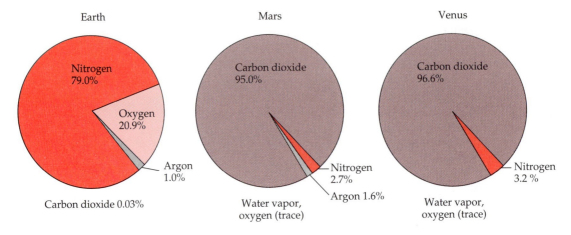

FIGURE 10

A comparison of the major components of the present atmospheres of Mars, Venus, and the earth. The percentages represent number of molecules, not relative weights. Elements without percentage values are present in trace amounts. (After Margulis and Olendzenski 1992.)

and return them unchanged, but alters their chemistry in ways that are beneficial to life. For example, were it not for the ammonia (a nitrogen–hydrogen compound, NH_3) that is produced in large amounts by organisms, the waters and soils of the earth would be so acidic that only a few organisms now on earth could survive. A variety of specialized microorganisms (e.g., nitrogen fixers and denitrifiers) play major roles in keeping this vital compound moving in an orderly manner between biotic and abiotic states. Without the critical buffering activities of early life forms and the continued coordinated activities of plants and microbes that dampen fluctuations in physical factors, conditions on earth, according to Lovelock and Margulis, would be similar to current conditions on Venus: very hot, with no oxygen in the atmosphere.

Controls in the Ecological Hierarchy

According to the Gaia hypothesis, the biosphere is a highly integrated and self-organized **cybernetic** (from *kybernetes*: pilot or governor) or controlled system. But cybernetics at the biosphere level is not accomplished by external, goal-oriented set-point controls like the thermostats, chemostats, or other mechanical feedback devices we use to control temperature and other conditions in our houses. Rather, control is internal and diffuse, involving hundreds of thousands of feedback loops and synergistic interactions in subsystems such as the microbial network that controls the nitrogen cycle. The contrast between these two types of controls in the ecological hierarchy is shown in Figure 11. From the organism level down, neural, hormonal, and genetic mechanisms maintain a tight control on growth and bodily functions. Above the organism level, control by negative and positive feedback is much less precise, so when limits are approached, there tends to be a pulsing state rather than a steady state. Waddington (1975) coined the term **homeorhesis** (from the Greek meaning "maintaining the flow") to denote evolutionary and ecological stability as opposed to **homeostasis**, the widely used term for physiological stability at the organism level. (For further discussion of differences between cybernetics at the individual and ecosystem levels, see Patten and Odum 1981, and for more about pulsing as an ecosystem characteristic, see W. E. Odum et al. 1995.)

Since we humans did not build the biosphere, we do not fully understand it, and as previously noted, we have not yet been able to construct even a simplified biologically controlled life-support system for space travel. We have much to learn about what really goes on in the impenetrable (to the naked eye) networks in the oceans and the "brown belts" of soils and sediments that determines when, where, and at what rate nutrients are recycled and gases exchanged. Lovelock admits that the "search for Gaia" (i.e., proof of the hypothesis) will be long and difficult, since so many processes would have to be involved in a control network of such magnitude.

Because of the absence of set-point controls, many scientists are skeptical that ecosystems and the biosphere really function as cybernetic systems, although most accept the concept that organisms play major roles in the control of the chemistry of the atmosphere and the oceans (Kerr 1988). The fact that catastrophic events, such as comets crashing into the earth, massive volcanic eruptions, and glaciers, have occurred from time to time raises questions about global homeostasis. Yet, despite a loss of species during these geological and cosmic upheavals, life has not only persisted but has continued to diversify and play a role in restoring favorable conditions for itself. However, just because the biosphere has exhibited the resilience stability to recover in past ages is no reason to be complacent about the resilience of our current life-support systems. Humans as a species might not survive a human-made catastrophe such as nuclear war or toxification of the oceans; and even if we did survive, all our hard-earned cultural and lifestyle gains would be wiped out.

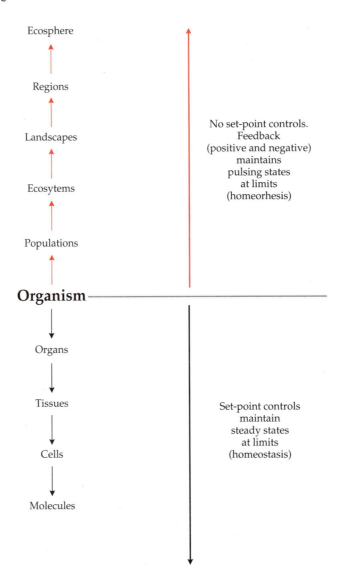

FIGURE 11

Cybernetics in the organism–ecosystem hierarchy. Compared with the strong set-point controls at the organism level and below, organization and function at the ecosystem level and above are much less tightly regulated, with more pulsing and chaotic behavior, but they are controlled nevertheless by alternating positive and negative feedback—in other words, they exhibit homeorhesis as opposed to homeostasis. Failure to recognize this difference in cybernetics has resulted in much confusion about the realities of the "balance of nature."

Land Without Life

The Copper Basin at Copperhill, Tennessee, shown in Figure 12, illustrates on a small scale what land without life would be like. Around the beginning of the twentieth century, sulfuric acid fumes from copper smelters (an extreme case of acid rain) exterminated all of the rooted plants over a sizable area. Most of the soil then eroded away, leaving a landscape that looked very much like that on Mars. Even though many years of court battles finally forced the industry to reduce the fumes, attempts to revegetate the area have been expensive and difficult. By bringing in topsoil, a thin cover of grass has been established on some of the eroded area. In some of the less severely eroded places some heavily fertilized pine plantations, well stocked with imported mycorrhizae, have been established. There is an even larger human-made desert near Sudbury, Ontario, Canada, created by the same kind of "roast-smeltering" as at Copperhill (in this case it was nickel, rather than copper, that was being mined). The area affected is so large that it is clearly visible from space as a scar on the face of the earth. A very large and expensive restoration program is now under way, in which it is planned to leave a large area unrestored as a comparison (Gunn 1995).

Even if the Gaia hypothesis cannot be verified at this time, we know enough to understand the importance of preventing pollution, not only

FIGURE 12
The Copperhill landscape in Tennessee as it looked in the 1930s, at the height of the release of acidic fumes by the copper smelters. (Photograph courtesy of U.S. Forest Service.)

> ## The Lessons of Copperhill
>
> Important economic and political lessons can be learned from Copperhill. When a single industry uses up all of the life-support capacity of an area and destroys, perhaps permanently, a good part of it, no further economic development in that area is possible. No other industry or business can come in, because no environmental support exists for anything else. People living in the area are mired in an unhealthy environment and the one-industry syndrome of political domination and cultural stagnation. Furthermore, little or no profit from an exploitative industry of this type remains in the area; the money goes to other areas where economic development is still possible. In 1988, the copper company at Copperhill announced it was closing down permanently. Perhaps the local communities can develop a tourist industry so people can see the "painted desert of Tennessee." The web of life on which our own life depends is tough and resilient, but once destroyed, it becomes extremely expensive, perhaps impossible, to restore within a human lifetime.

the severe local type as seen at Copperhill or at a toxic waste dump, but also the much more difficult problem of widespread, or nonpoint, sources of pollution, which can severely stress the intricate bioregenerative network that keeps our planet livable. We will look at some of these nonpoint threats in subsequent chapters.

In a recent article, Serafin (1988) analyzed the two extreme views on how the earth is controlled: Vernadsky's "noösphere" (1945), or the domination of earth by the mind of man, and Lovelock's Gaia, or the control of earth by organisms other than humans. Serafin suggested that the two concepts are merging to form a new science of the biosphere that gives credence to both concepts.

Suggested Readings

*Bailey, R. G. 1995. *Descriptions of ecoregions of the United States*, 2nd ed. Forest Service, U.S. Dept. Agriculture Misc. Publ. 1391, Fort Collins, CO.

*Barrett, G. W. 1968. The effects of an acute insecticide stress on a semi-enclosed grassland ecosystem. *Ecology* 49: 1019–1035.

*Benner, R., J. D. Pakulski, M. McCarthy, J. I. Hedges, and P. G. Hatcher. 1992. Bulk chemical characteristics of dissolved organic matter in the ocean. *Science* 255: 1561–1565.

*Brown, J. H., D. W. Mehlman, and G. S. Stevens. 1995. Spatial variation in abundance. *Ecology* 76: 2028–2043.

*Carson, R. 1962. *Silent Spring*. Houghton Mifflin, Boston.

*Chapin, F. S., E. D. Schulze, and H. A. Mooney. 1992. Biodiversity and ecosystem processes. *Trends Ecol. Evol.* 7: 107 -108.

Cloud, P. E. 1988. Gaia modified. *Science* 240: 1716.

Cole, J., G. Lovett, and S. Findley, eds. 1991. *Comparative Analysis of Ecosystems*. Springer-Verlag, New York.

*Decamps, H., and R. J. Naiman. 1990. Towards an ecosystem perspective. Chapter 1 in *The Ecology and Management of Aquatic–Terrestrial Ecotones*, eds. R. J. Naiman and H. Decamps. Man and the Biosphere Series, vol. 4. Parthenon Publishing Group, Park Ridge, NJ.

Evans, F. C. 1956. Ecosystem as the basic unit in ecology. *Science* 123: 1127–1128.

*Feinsinger, P. 1994. Habitat "shredding." In *Conservation Biology*, ed. G. Meffe and R. Carroll, pp. 258–260. Sinauer Associates, Sunderland, MA.

*Forman, R. T. T. 1995a. *Land Mosaics: The Ecology of Landscapes and Regions*. Cambridge University Press, Cambridge.

*Forman, R. T. T. 1995b. Some general principles of landscape and regional ecology. *Landscape Ecol.* 10: 133–142.

Golley, F. B. 1993. *A History of the Ecosystem Concept in Ecology*. Yale University Press, New Haven, CT.

*Gunn, J., ed. 1995. *Restoration and Recovery of an Industrial Region: Progress in Restoring the Smelter-Damaged Landscape near Sudbury, Canada*. Springer Verlag, NY.

Hagen, J. B. 1991. *An Entangled Bank: The Origin of Ecosystem Ecology*. Rutgers University Press, New Brunswick, NJ. (A goal of ecosystem ecology is to untangle Darwin's "tangled bank" commentary in *The Origin of Species*.)

*Harris, L. D. 1984. *The Fragmented Forest*. University of Chicago Press, Chicago.

Holland, M. M., P. G. Risser, and R. J. Naiman, eds. 1991. *Ecotones: The Role of Landscape Boundaries in the Management and Restoration of Changing Environments*. Chapman and Hall, New York.

*Hubbell, S. P. 1979. Tree dispersion, abundance, and diversity in a tropical dry forest. *Science* 203: 1299–1309.

*Janzen, D. 1987. Habitat sharpening. *Oikos* 48: 3–4.

*Jordan, L. L. 1981. The birth of ecology: An account of Alexander von Humboldt's voyage to the equatorial regions of the new continent. In *Tropical Ecology*, ed. C. F. Jordan, pp. 4–15. Benchmark Papers in Ecology, 10. Hutchinson-Ross Publishing, Stroudsburg, PA.

*Kerr, R. A. 1988. No longer willful, Gaia becomes respectable. *Science* 240: 393–395.

Krall, F. R. 1994. *Ecotone: Wayfaring on the Margins*. State University of New York Press.

*Kruess, A., and T. Tscharntke. 1994. Habitat fragmentation, species loss, and biological control. *Science* 264: 1581–1584.

Lawson, G. J., G. Cottam, and O. L. Loucks. Unpublished manuscript. Terrestrial primary production of adjacent urban and natural ecosystems.

*Lovelock, J. E. 1979. *Gaia: A New Look at Life on Earth*. Oxford University Press, New York.

Lovelock, J. E. 1988. *The Ages of Gaia: A Biography of Our Living Earth*. Norton, New York.

*Mann, K. H. 1973. Seaweeds: Their productivity and strategy for growth. *Science* 182: 975–981.

Margulis, L. 1982. *Early Life*. Science Books International, Boston.

*Margulis, L., and L. Olendzenski, eds. 1992. *Environmental Evolution: Effects of the Origin and Evolution of Life on Planet Earth*. M.I.T. Press, Cambridge, MA.

*Naveh, Z., and A. S. Lieberman. 1984. *Landscape Ecology Theory and Application*. Springer Verlag, New York.

*Odum, E. P. 1982. Diversity and the forest ecosystem. *In Natural Diversity in Forest Ecosystems*, J. L. Cooley and J. Cooley, eds. pp. 35–41. U.S.D.A. Forest Service Southeastern Experiment Station, Asheville, NC.

Odum, E. P. 1986. Introductory review: Perspectives on ecosystem theory and application. In *Ecosystem Theory and Application*, ed. N. Polunin, pp. 1–11. John Wiley & Sons, New York.

*Odum, E. P., and R. L. Kroodsma. 1977. The power park concept: Ameliorating man's disorder with nature's order. In *Thermal Ecology*, vol. II, eds. G. W. Esch and R. W. McFarlane. National Technical Information Service, Springfield, VA.

Odum, H. T. 1971. *Environment, Power, and Society*. Wiley-Interscience, New York.

Odum, H. T., and E. C. Odum. 1981. *Energy Basis for Man and Nature*, 2nd ed. McGraw-Hill, New York. (See pp. 293–294 for energy language symbols.)

Odum, W. E. 1990. Internal processes influencing the maintenance of ecotones: Do they exist? Chapter 6 in *The Ecology and Management of Aquatic–Terrestrial Ecotones*, eds. R. J. Naiman and H. Decamps, pp. 91–102. Parthenon Publishing Group, Park Ridge, NJ.

*Odum, W. E., E. P. Odum, and H. T. Odum. 1995. Nature's pulsing paradigm. *Estuaries* 18: 547–555.

Omernik, J. M. 1987. Ecoregions of the conterminous United States. *Ann. Assoc. Am. Geogr.* 77: 118–125.

*Patten, B. C., and E. P. Odum. 1981. The cybernetic nature of ecosystems. *Am. Nat.* 118: 886–895.

Platt, A. 1995. Dying seas. *Worldwatch* 8(1): 10–19.

Polunin, N., ed. 1985. *Ecosystem Theory and Application*. John Wiley & Sons, New York.

Pomeroy, L. R., and J. J. Alberts, eds. 1988. *Concepts of Ecosystem Ecology: A Comparative Review*. Springer-Verlag, New York.

*Redfield, A. C. 1958. The biological control of chemical factors in the environment. *Am. Sci.* 46: 205–221.

*Rhoads, R. E. 1990. The world's food supply at risk. *Nat. Geog.* 179(4): 74–105.

*Rice, E. L. 1952. Phytosociological analysis of a tall-grass prairie in Marshall County, Oklahoma. *Ecology* 33: 112–116.

*Sachs, A. 1995. Humboldt's legacy and the restoration of science. *Worldwatch* 8(2): 28–38.

*Schindler, D. W. 1990. Experimental perturbations of whole lakes and tests of hypotheses concerning ecosystem structure and function. *Oikos* 57: 25–41.

*Serafin, R. 1988. Noosphere, Gaia and the science of the biosphere. *Environ. Ethics* 10: 121–137.

*Sousa, W. P. 1984. The role of disturbance in natural communities. *Annu. Rev. Ecol. Syst.* 15: 353–391.

*Tansley, A. G. 1935. The use and abuse of vegetational concepts and terms. *Ecology* 16: 284–307.

*Tilman, D., D. Widen, and J. Knops. 1996. Productivity and sustainability in grassland ecosystems. *Nature* 379: 718–720.

*Turner, M. G. 1989. Landscape ecology: The effect of pattern and process. *Annu. Rev. Ecol. Syst.* 20: 171–197.

Vernadsky, V. I. 1929. *La Biosphere*. Nouvelle Collection Scientifique. Felix Alcan, Paris, France. (English translation published 1986, Synergetic Press, London.)

*Vernadsky, V. I. 1945. The biosphere and the noösphere. *Am. Sci.* 33: 1–12.

*Waddington, C. H. 1975. A catastrophe theory of evolution. In *The Evolution of an Evolutionist*, pp. 153–266. Cornell University Press, Ithaca, NY.

*Whitcomb, R. F., C. S. Robbins, J. F. Lynch, B. L. Whitcomb, M. K. Klimkiewicz, and D. Bystrak. 1981. Effects of forest fragmentation on avifauna of the eastern deciduous forest. Chapter 8 in *Forest Island Dynamics in Man-Dominated Landscapes*, eds. R. L. Burgess and D. M. Sharp. Springer-Verlag, New York.

Wiegert, R. G., and D. F. Owen. 1971. Trophic structure, available resources and population density in terrestrial and aquatic ecosystems. *J. Theor. Biol.* 30: 69–81. (Contrasts structure and function of terrestrial and aquatic ecosystems.)

*Wilcox, B. A. 1984. In situ conservation of genetic resources: Determinants of minimum area requirements. In *National Parks: Conservation and Development*, eds. J. A. Neeley and K. R. Miller, pp. 639–647. Smithsonian Institution Press, Washington, D.C.

*Wilson, E. O., ed. 1988. *Biodiversity*. National Academy Press, Washington, D.C. (38 articles by different authors on the preservation, value, and restoration of diversity.)

*Zonneveld, I. S., and R. T. T. Forman, eds. 1990. *Changing Landscapes: An Ecological Perspective*. Springer-Verlag, New York.

Of the many well-written magazine treatments of the state of our environment that have appeared recently, the following special issues are especially recommended:

Time 133(1): 24–73, January 2, 1989. *Planet of the Year: Endangered Earth.*
Scientific American 261(3), September 1989. *Managing Planet Earth.*

*Indicates references cited in this chapter.

4

Energetics

If one were asked to pick out a single common denominator of life on earth, that is, something that is absolutely essential and involved in every action, large or small, the answer would have to be energy. In physics books, as well as in the dictionary, **energy** is defined as the ability or capacity to do work, with *work* defined in its broadest sense of "to do or perform something." While you are working at your job or hobby, or relaxing, or even sleeping, your body is performing thousands of vital functions that require energy of a specific amount and kind. The fantastic variety of organisms and functions involved in keeping our life-support systems operating require a great deal of energy. The primary energy source for heterotrophs, of course, is food; for autotrophs it is light and the indirect solar energies (wind, rain) required for photosynthesis. In addition, human societies, especially industrialized ones, require large amounts of concentrated energy in the form of fuels. All humans should understand the basic principles of energy transformations, because without energy, there can be no life. We should start teaching energetics in the first grade.

In 1976 Barry Commoner wrote about (1) our failure to understand the interlocking of three systems—ecosystem, production system, and economic system—linked by energetics and (2) the futility of "quick-fixing" a problem in any one of these systems as if it were separate from the others. But it has been H. T. Odum who has developed most fully the concept of energy as the common denominator in ecology and the link between ecosystem and economic system (H. T. Odum 1971, 1973, 1996; Odum and Odum 1981).

Energy Units

Unfortunately, there are many different units that are widely used to characterize quantities of energy. We have the calorie, the watt, the joule, the erg, the BTU, and the horsepower, to mention a few, each of which was originally developed or used to measure a particular kind of energy (e.g., watts for electricity, calories for food, and barrels for oil). While all these units are roughly interconvertible, it almost takes a computer to calculate the total energy input into your home, since electricity, heat, gasoline, oil, and gas are all reported on your bills in different units. To add to the confusion, some units, such as the calorie, represent potential energy without respect to time, while other units, such as the watt, are rate-of-power units with time built into the definition. This chaotic situation results from excessive fragmentation of disciplines (each of which clings to its traditional usage) and lack of communication and coordination between industries and nations. A concerted effort is now being made to designate one unit (perhaps the joule or the watt) as the international unit to be used by all nations for all energies.

To keep things as simple as possible in this book, we will use the **calorie** as our energy quantity unit, since most people are already somewhat familiar with it in connection with our food intake ("only one calorie per bottle," as soft drink advertisers chant). We will use the small or gram calorie (c), abbreviated **gcal**, and the large or kilogram Calorie (C), abbreviated **kcal**. One gcal is equivalent to the heat necessary to raise the temperature of one gram of water one degree centigrade. The kilocalorie is 1000 times larger, equivalent to the heat necessary to raise the temperature of one kilogram of water one degree centigrade. In nutrition, as well as in ecology, the kilocalorie is usually used. To provide a reference point for the numbers we will use to compare various energy flows, remember that it takes about 2000–3000 kcal per day, or roughly a million kcal per year, of food of the proper quality to power your body.

As already noted, the joule and the watt are proposed as international units to be used for all kinds of energies. A **joule** is defined as the amount of work energy required to raise one kilogram to the height of 10 centimeters (about 0.24 gcal); a **watt** is defined as one joule per second. Since the watt and joule are such small amounts of energy, the **kilowatt-hour** (kWh = 1000 watts/hour) is widely used, and is equal to about 860 kcal. This, of course, is the unit that appears on your electric bill. The **kilojoule**

(KJ = 1000 joules) is also widely used as an international unit in ecology; it equals 4.2 kcal (about the same as a BTU).

While there are too many quantitative units, strangely enough there is no generally accepted unit for expressing **energy concentration**, or the "quality" of different types of energy, which differ greatly in their ability to do a given kind of work. For example, 100 Calories of sunshine is not at all equal to 100 Calories of gasoline when it comes to running your car. We will see what is being done about this oversight in the section on energy concentration below.

Energy Laws

The behavior of energy is governed by two laws, known as the **laws of thermodynamics**. The first law states that energy may be transformed from one form (such as light) into another (such as food), but is never created or destroyed. The second law states that no process involving an energy transformation will occur unless there is a degradation of energy from a concentrated form (such as food or gasoline) into a dispersed form (such as heat). Because some energy is always dispersed into unavailable heat energy, no spontaneous transformation (light to food, for example) can be 100 percent efficient.

The second law is sometimes known as the entropy law. **Entropy** (from *en*: in; *trope*: transformation) is a measure of **disorder** in terms of the amount of unavailable energy in a closed thermodynamic system. Thus, although energy is neither created nor destroyed during transformation, some of it is degraded to an unavailable or less available form (dispersed heat, for example) when used (i.e., transformed). The food you ate for breakfast is no longer available to you once it has been respired in the maintenance of your body tissues; you must go to the store and buy more for tomorrow. *Energy cannot be reused*, in contrast with materials such as water, nutrients, or money, which can be recycled and used over and over again with little or no loss of utility. The two energy laws are illustrated in the diagram of energy flow through an oak leaf in Figure 1. The general picture of energy inflows, transformations and storages, and outflows at the ecosystem level was shown in our first models, Figures 1 and 2 in Chapter 3.

Organisms and ecosystems maintain their highly organized, low-entropy (low-disorder) state by transforming energy from high to low utility states. Living systems and the whole biosphere are what Ilya Prigogine has called "far-from-equilibrium systems" that have efficient "dissipative structures" to pump out the disorder (Prigogine et al. 1972). Pri-

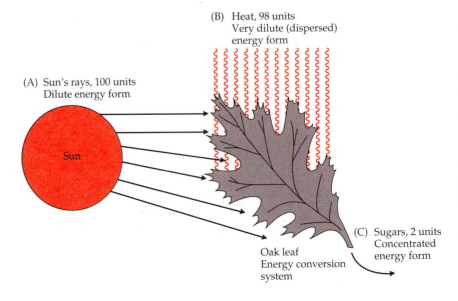

FIGURE 1
The two laws of thermodynamics. The first law is illustrated by the conversion of sun energy (A) to food (sugar, C) by photosynthesis (A = B + C). The second law dictates that (C) is always less than (A) because of heat dissipation (B) during conversion.

gogine won a Nobel Prize for his theoretical analysis of nonequilibrium thermodynamics. He was able to explain better than anyone before him how living systems seemingly defy the second law by self-organizing to maintain an open, far-from-equilibrium state. Entropy, it turns out, is not all negative; as the quantity of energy declines in successive transfers, the quality of the remainder may be greatly enhanced.

The respiration of the highly ordered biotic community is the dissipative structure of an ecosystem. By analogy, well-organized maintenance crews and plenty of tax dollars are required to dissipate disorder and maintain quality in a city. If the quality and quantity of energy flow through a forest or a city is reduced, or if disorder dissipation is inadequate, then the forest or the city begins to degrade and senesce (become disorderly). This happens to a lot of cities, and to forests that are unduly stressed by air pollution.

To survive and prosper, natural and human-made ecosystems alike require a continuous input of high-quality energy, storage capacity (for periods when input is less than needed), and the means to dissipate entropy. These three attributes are part of what H. T. Odum calls the **maximum power principle**, which states that the systems most likely to survive in this competitive world are those that efficiently transform the most energy into useful work for themselves and surrounding systems with which they are linked for mutual benefit (Odum and Odum 1981).

The person who introduced thermodynamics to ecology was Alfred James Lotka (1880–1949), a physical chemist by training. While working for a chemical company, he devoted his spare time to developing a new field that he called "physical biology." Lotka's basic thesis was that the organic and inorganic world functioned as a single system with all components linked through thermodynamics in such an intimate way that it was impossible to understand the part without understanding the whole. In 1925 he published a book entitled *Elements of Physical Biology*, outlining his theories developed over a period of 20 years. His book, which was to become a classic (republished in 1956 under the title *Elements of Mathematical Biology*), attracted the attention of ecologist Charles C. Adams, who convinced him to join the Ecological Society of America, and the famous demographer Raymond Pearl, who arranged an appointment for him at Johns Hopkins University so he could continue his studies in an academic environment. Here Lotka was to make major contributions to mathematical models of populations (which will be discussed in Chapter 6).

Lotka felt that biologists took too narrow a view of evolution when they focused their attention on individual species. He believed that natural selection must also operate on the energy flow of the undivided system; this concept is now known as the maximum power principle. It is significant that a biologist (Tansley) and a physical scientist (Lotka) independently came up with the idea of the ecological system as a major functional unit in the biosphere. Because he coined the word "ecosystem" and it has caught on, Tansley gets most of the credit, which should be shared with Lotka.

Solar Radiation

Sunlight is one of those familiar blessings that we take for granted without bothering to know much about it ("familiarity breeds contempt," as the old saying goes). For one thing, you can see only about half of what comes in from the sun every day. To put the sun's rays into perspective, the entire spectrum of electromagnetic (wave-type energy) radiation is shown in Figure 2. Solar radiation is in the middle range of this spectrum, with wavelengths largely between 0.1 and 10 micrometers (1 micrometer = 1/1000 millimeter). Solar radiation consists of visible light and two invisible components, ultraviolet light and infrared light. The long-waved

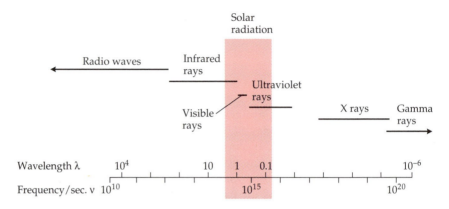

FIGURE 2
The spectrum of electromagnetic radiation. Solar radiation is in the middle range of the spectrum.

infrared radiation is the "heating" part of sunlight. The visible range is not only the light we see by, but is the energy used in photosynthesis (which is restricted to the visible light). Much of the ultraviolet light reaching the upper atmosphere is kept out by the **ozone shield**, which is fortunate since ultraviolet light is lethal to exposed protoplasm. Ozone, a molecule that consists of three oxygen atoms (O_3), is formed in the upper atmosphere (the stratosphere) by the interaction of oxygen and ultraviolet radiation. The ozone shield developed early in the history of the biosphere as oxygen began to accumulate; the shield made it possible for primitive life to emerge from the waters and evolve into the higher forms present today.

The absorption of solar radiation by the atmosphere as it enters the biosphere greatly reduces ultraviolet light, broadly reduces visible light, and irregularly reduces infrared light (some wavelengths are reduced more than others). Radiant energy reaching the surface of the earth on a clear day is about 10 percent ultraviolet, 45 percent visible, and 45 percent infrared. The visible radiation is the least attenuated as it passes through cloud cover and water, which means that photosynthesis can continue on a cloudy day (though often at reduced rates) and at some depth in clear water. Vegetation absorbs the blue and red visible wavelengths (the most useful for photosynthesis) and the far-infrared strongly, the green less strongly, and the near-infrared very weakly. By rejecting the near-infrared, in which the bulk of the sun's heat energy is located, leaves of terrestrial plants avoid lethal temperatures (they are also cooled by evaporation of water). This interfacing of the sun's rays and the earth's green mantle is another example of how organisms not only adapt to physical conditions but also modify them to suit their needs. The cool, green shade of a forest, due to the absorption of red and blue visible light and near-infrared light by the foliage overhead, is indeed a different radiation environment from human-made deserts such as the Copperhill landscape or a parking lot. There are thousands of kinds of animals and microbes living in the forest, but probably only a few transients (including humans) to be found in the parking lot, all probably trying to get to a better place.

Because the green light and the near-infrared light are reflected by vegetation, these spectral bands are used in aerial and satellite remote sensing and photography to reveal patterns of natural vegetation, the conditions of crops, the presence of diseased plants, and so on. The continuing development of the technology of **remote sensing** and **geographic information systems** (computer renditions and analyses of remote sensing imagery) is one of the more positive benefits of the space age.

Threats to the Ozone Shield

Certain chemicals produced in manufacturing, such as the chloroflurocarbons (CFCs) used as aerosol propellants (Cohn 1987) and emissions from high-flying jet aircraft, break down ozone. Ozone-destroying substances have been getting into the upper atmosphere in increasing amounts, to the point where they pose a serious threat to the ozone shield. The first warning that the shield might be thinning over the Antarctic came in 1974, and what has come to be called the **ozone hole** first appeared over that region 1984. By 1991 this hole had enlarged to include southern Argentina, where increases in skin cancers and blindness in sheep and humans have been reported. Accumulation of ozone-destroying chlorine monoxide over Europe indicates drastic thinning of the shield in the temperate zone. In 1987, the industrial nations signed what has come to be known as the Montreal Protocol, agreeing to reduce the production of CFCs in the hope of halting further damage to the life-protecting shield. By 1996 the agreement seemed to be working, since the production of ozone-destroying chemicals had decreased as industries developed substitutes (Kerr 1996). (Ozone as a pollutant at ground level will be discussed in Chapter 5.)

The Radiation Environment at Ground Level

Organisms at or near the surface of the earth are immersed in a radiation environment consisting not only of direct downward-flowing solar rays, but also long-wave heat (thermal) radiation from nearby surfaces. Both components contribute to temperature and other conditions of existence, as outlined in Chapter 5. Thermal radiation comes not only from soil, water, and vegetation, but also from clouds, which radiate a substantial amount of heat energy downward into ecosystems. You may have observed that the temperature on a winter night remains higher when it is cloudy than when it is clear; it's the reradiated heat from clouds that makes the difference. Even colorless gases such as carbon dioxide trap heat and reflect it downward, creating a "greenhouse effect" (so called because carbon dioxide, like the glass in a greenhouse, allows incoming solar rays to pass through, but reflects outgoing thermal radiation). In addition to worrying about ozone depletion, we also have to worry about the fact that atmospheric carbon dioxide is increasing due to human activities, which could affect our climate. We will take up this worry in Chapter 5.

Energy Flow through the Biosphere

Figure 3 and Table 1 show what happens to solar radiation as it passes through the biosphere, performing useful work every step of the way. On a square meter basis, solar energy comes in at the rate of about 5 million kcal per year. This large flow is reduced exponentially as it passes through the clouds, water vapor, and other gases of the atmosphere, so the annual amount actually reaching the autotrophic layer of ecosystems is only about 1–2 million kcal per square meter (lowest in the cloudy north and highest in the deserts). Of this amount, about half is absorbed

TABLE 1 Energy Dissipation of Solar Radiation as Percentage of Annual Input into the Biosphere	
Energy Dissipation	**Percent**
Reflected	30
Direct conversion to heat	46
Evaporation, precipitation (drives hydrological cycle)	23
Wind, waves, and currents	0.2
Photosynthesis	0.8
Total	100

Tidal energy: about 0.0017%
Terrestrial heat: about 0.5%

Data from Hulbert 1971.

(A) Pictorial diagram

(B) Energy flow diagram

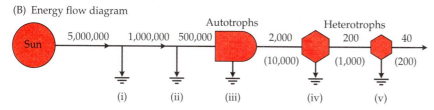

FIGURE 3

Solar energy flow through the biological food chain, in kcal per square meter per year. (A) is a pictorial diagram; (B) is a more formal energy flow model. Figures in parentheses show levels that may be reached in subsidized ecosystems when solar energy flow is enhanced by other types of energy such as fuels.

by a well-stocked green layer, and about 1 percent of this on average (up to 5 percent under the most favorable conditions) is converted to organic matter by photosynthesis.

As shown in Figures 1 and 3, a large part of the solar energy flow is dissipated into unavailable heat at each transfer, as required by the second law of thermodynamics. This dissipation is not wasted energy because useful work is accomplished at each transfer, not just in the biological part, but all along the chain. For example, the dissipation of solar radiation as it passes into the atmosphere, the oceans, and the greenbelts warms the biosphere to life-sustaining levels, drives the hydrological cycle (evaporates water, which returns as rain), and powers weather systems. An estimate of the percentage of solar radiation involved in major transfers is shown in Table 1. Note that about a quarter of the solar energy flow is used to recycle water, one of the biosphere's most vital nonmarket services.

It is the flow of energy that drives the cycles of materials. To recycle water and nutrients requires an expenditure of energy, which is not recyclable—a fact not understood by those who think that artificial recycling of resources such as water, metals, or paper is somehow an instant and free solution to shortages. Like everything else worthwhile in this world, there is an energy cost for this (and if done artificially there is also a monetary cost).

Nature certainly makes good use of solar power. To what extent humans can tap this source to replace fossil fuels as they run out remains to be seen. One advantage of solar energy compared with fossil fuels is that it is renewable: using it does not deplete the source. On the other hand, solar energy is much more dilute (less concentrated) and won't run automobiles or machinery directly. It has to be converted to electricity or some kind of fuel to do this kind of work, and, like any transformation, this conversion has an entropy cost. There is no way we can get around the energy laws; recognizing such limitations is not meant to be discouraging, but to be realistic.

Energy Concentration: eMergy

We have already prepared the ground for the principle that as energy is used and dispersed in a chain or hierarchy of successive transformations, it changes in form and becomes increasingly concentrated or very high in information content. In other words, as energy quantity decreases, its "quality" increases. Not all Calories (or whatever quantitative unit is employed) are equal, because the same quantity of different forms of energy varies widely in work potential. As noted in the preceding paragraph, highly concentrated forms such as oil have a higher work potential and, therefore, a higher quality than do dilute forms such as sunlight. We can express energy concentration or quality in terms of the amount of one type of energy required to develop a certain amount of another type. The term **eMergy** (spelled with a capital *M*) has been proposed for this measure, and is defined as the available energy already used directly or indirectly to create a service or product (H. T. Odum 1996). Thus, if 1000 Calories of sunlight is converted to 1 Calorie of food by plants, the transformation ratio (or **transformity**) is 1000 sun Calories to 1 food Calorie, and the eMergy of the food is 1000 solar energy Calories. EMergy can be thought of as "energy memory" since it is calculated by adding up all of the energies at different levels that combined to produce the product or service. For comparative purposes, all the energies should be of the same kind (solar energy, for example). The term **embodied energy** has also been used for this kind of quality measure.

Increasing energy concentration with decreasing energy quantity is illustrated by the flow diagrams in Figure 4. In the natural food chain (Figure 4A), the amount of energy declines with each step, but concentra-

(A) Food chain

(B) Electric energy chain

FIGURE 4
Increasing energy concentration (quality) accompanies decreasing quantity in food chains and in electric energy generation. (After H. T. Odum 1983.)

tion in terms of the number of solar kilocalories dissipated increases. An estimated 10,000 kcal of sunlight are estimated to be required to produce 1 kcal of predator. 100 energy units of herbivores are required for every one of predators. Accordingly, predators are relatively rare and energy-expensive components of ecosystems, but they may be very important in terms of feedback control of herbivores, which in turn may have a major effect on plant production.

Figure 4B shows an energy chain leading to the generation of electricity. (It must be emphasized that the numbers on the charts are tentative, rounded-off "orders of magnitude," subject to revision as more attention is devoted to means of comparing different types of energy.) Again, as the quantity declines along the chain, the capacity to perform work per unit of quantity (Calorie) increases with each conversion. Thus, the energy of the sun is concentrated by green plants, further concentrated by the fossilization process that produces coal, and finally enhanced some more in the production of electricity, which, accordingly, is a form of energy derived from solar energy but concentrated many thousands of times.

The production of electricity, whether by burning fossil fuels or by direct conversion with solar cells, uses up a lot of energy, but its utility is so great (think of all the tasks it can perform) that it would be difficult to imagine our world without it.

In summary, eMergy and embodied energy are measures of the worth or value of a product or service, since they represent how much energy of one kind had to be "paid" to produce another, more "valuable," form. The point is that the amount of energy directly involved in a process or contained in a product is not at all a measure of its actual energy cost. For example, thinking or reading this book requires very little energy, hardly a small calorie (gcal) an hour. However, since a great deal of energy is used to develop the ability to read and think, the embodied energy or eMergy of this activity is large. Education, and intellectual activity in general, is thus very energy-expensive—which helps to explain why it costs so much to support quality education for large numbers of people. (We will discuss energy–money conversions later in this chapter.)

Primary Production

The rest of this chapter is devoted to that part of the solar energy flow that is converted to food and other high-quality organic matter, since it is upon this that the entire living world depends. The 1–5 percent conversion efficiency previously noted would seem to be very low in view of the much higher energy conversion efficiencies of electric motors and other machines. Actually, photosynthesis and engine efficiency are not directly comparable, because the energy source for engines is of a much higher quality (more concentrated) than solar energy. Also, the large amount of energy used in making, repairing, and replacing a machine is not included in calculating its efficiency, while self-maintenance and self-perpetuation are part of the energy budget in a biological system. Photosynthesis turns out to be a very efficient process for tapping that small portion of solar radiation that can be upgraded to high-utility organic matter. No one has been able to improve on this aspect of nature's performance. To be sure, we get more food from plants these days, but not by increasing the efficiency of photosynthesis. What we do is increase the amount of conversion that ends up in a harvestable form that we (or our domestic animals) can eat. We'll see how this is done shortly.

Primary production and **primary productivity** are terms used to designate the amount of organic matter fixed (converted from solar energy) by autotrophs in a given area over a given period of time, generally expressed as a rate, so much per day or year. **Gross primary production** is the total amount, including that used by a plant for its own needs, while

net primary production is the amount stored in a plant in excess of its respiratory needs and therefore potentially available to heterotrophs. **Net community production** is the amount left after the biotic community, autotrophs and heterotrophs, have taken all the food they need. Finally, energy storage at consumer levels, in cows or fish, for example, is referred to as **secondary production**.

High rates of primary production in both natural and cultured ecosystems occur when physical factors (e.g., water, nutrients, and climate) are favorable, and especially when auxiliary energy from outside the system reduces maintenance costs (i.e., enhances disorder dissipation). Secondary energy that supplements the sun and allows a plant to store and pass on more of its photosynthate can be thought of as an **energy subsidy**. Wind and rain in a rain forest, tidal energy in an estuary, or fossil fuel used in cultivating a crop are examples; all of these enhance production by plants and animals adapted to make use of the auxiliary energy. For example, tides do the work of bringing nutrients to marsh grasses and food to oysters as well as taking away waste products, so the organisms do not have to expend energy for these jobs and can use more of their own production for growth.

Source–Sink Energetics

A corollary to energy subsidy is the concept of **source–sink energetics**, in which excess organic production by one ecosystem (a source) is exported to another, less productive, ecosystem (a sink). For example, a productive estuary may export organic matter or organisms to less productive coastal waters. Accordingly, the productivity of an ecosystem is determined by the rate of production within it plus that received as an import. At the species level, one population may produce more offspring than are needed to maintain it, with the surplus moving to an adjacent population that otherwise would not be self-sustaining (Pulliam 1988).

Kinds of Photosynthesis

The basic photosynthetic process is chemically an oxidation–reduction reaction, the equation for which can be written in word form as follows:

**carbon dioxide + water + light energy =
carbohydrate + water + oxygen**

It is water that is oxidized and carbon dioxide that is reduced (fixed) to carbohydrate, which leads to formation of other types of food and organic matter.

In most plants, carbon dioxide fixation starts with the formation of three-carbon compounds, but certain plants reduce carbon dioxide in a different manner, starting with four-carbon carboxylic acids. These two types of plants are called **C_3 plants** and **C_4 plants**. C_4 plants have a different arrangement of chloroplasts within their leaves and, more important, they respond differently to light, temperature, and water.

Figure 5 contrasts the responses of C_3 and C_4 plants to light and temperature. C_3 plants tend to peak in photosynthetic rate (per unit of leaf surface) at moderate light intensities and temperatures, and to be inhibited by high temperatures and the intensity of full sunlight. In contrast, C_4 plants are adapted to high light and temperature conditions, and use water more efficiently under these conditions. As would be expected, C_4 plants dominate desert and grassland communities in warm temperate

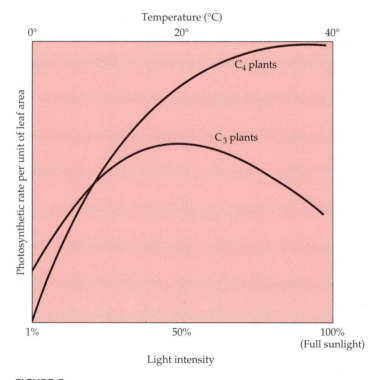

FIGURE 5
Comparative photosynthetic responses of C_3 and C_4 plants to increasing light intensity and temperature.

and tropical climates and are rare in forests and in the cloudy north where light intensities and temperatures are low to moderate. Despite their lower photosynthetic efficiency at the leaf level, C_3 plants account for most of the world's primary production, presumably because they are more competitive in communities of mixed species where light, temperature, and so on are average rather than extreme. Survival of the fittest in the real world does not always go to the species that is physiologically superior under ideal conditions in a monoculture; survival in the mixed culture of nature goes to those species that can perform and reproduce when conditions are not always the most favorable.

Major human food plants such as wheat, rice, and potatoes are C_3 plants, as are most vegetables. Crops of tropical origin such as corn, sorghum, and sugarcane are C_4 plants; so is Bermuda grass, which is widely used in southern pastures and golf courses. We might do well to domesticate more C_4 species for use in irrigated deserts and in the tropics.

There are other photosynthetic variants that are adapted to harsh conditions. Certain photosynthetic bacteria that can live under anaerobic conditions oxidize inorganic compounds such as hydrogen sulfide (H_2S) rather than water, and thus do not release oxygen. Succulent desert plants such as cacti have a photosynthetic mode known as **CAM** (crassulacean acid metabolism), in which carbon dioxide is absorbed and stored as organic acids at night but is not "fixed" until the next day when light becomes available, allowing the plant to keep its leaf openings (stomata) closed during the hot, dry daytime, thereby conserving water. There are probably many other as yet undiscovered adaptations for capturing the sun's energy no matter what the physical environment is like. (Nature has no end of schemes for getting along, which is one reason the study of nature is so fascinating!)

World Distribution of Primary Production

The general pattern of world distribution of primary production is shown in Figure 6. The range is very great (Whittaker and Likens 1971). Large parts of the open ocean and land deserts have an annual production rate of 1000 kcal per square meter or less. The ocean is nutrient-limited and, of course, deserts are water-limited. Many grasslands, coastal seas, shallow lakes, and ordinary agricultural communities range between 1000 and 10,000 kcal per square meter. Certain estuaries, coral reefs, moist forests, wetlands, and intensive (industrialized) agricultural and natural communities on fertile plains have annual production rates ranging from 10,000 to 25,000 kcal per square meter. In terms of area, about three-fourths of

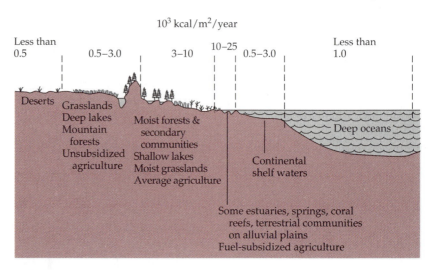

10^3 kcal/m^2/year

| Less than 0.5 | 0.5–3.0 | 3–10 | 10–25 | 0.5–3.0 | Less than 1.0 |

Deserts
Grasslands
Deep lakes
Mountain
forests
Unsubsidized
agriculture

Moist forests &
secondary
communities
Shallow lakes
Moist grasslands
Average agriculture

Deep oceans

Continental
shelf waters

Some estuaries, springs, coral
reefs, terrestrial communities
on alluvial plains
Fuel-subsidized agriculture

FIGURE 6
World distribution of primary production.

the biosphere is ocean and desert, and only about 10 percent is naturally very fertile. However, because of the large area involved, the total productivity of the less fertile regions is very large. And it is the energy flow through these regions that provides much of our life-support goods and services other than food.

Food for Humans

Human food production per unit of area has been greatly increased in recent decades by increasing mechanization, fertilizers, irrigation, and pesticides (Figure 7). All of these subsidies are energy inputs just as is sunlight; they can be measured in the calories or horsepower diverted to heat in the performance of the work of crop maintenance (the energy required to produce fertilizers and pesticides—their embodied energy—would be a measure of their contribution). The general relationship between crop yields and inputs of fertilizer, pesticides, and work energy is shown in Figure 8. Doubling the crop yield in the past has required about a tenfold increase in all these inputs. Increasing efficiency can reduce this high cost.

Genetic selection for increased **harvest ratio** (that is, ratio of edible to nonedible part of the crop plant) is another way yields have been increased. Recent success in breeding high-yielding varieties of cereal

FIGURE 7
Pesticide spraying of crops is an energy input that has increased food production, but has caused other problems as human food becomes increasingly contaminated and poisons spread far beyond the crop field. These problems can be corrected by less wasteful, more efficient integrated pest management. (Photograph by B. J. Miller/Biological Photo Service.)

grains has been called the **Green Revolution**. A wild rice plant puts no more than 20 percent of its production into seeds, enough to insure its long-term propagation; in contrast, highly bred strains of "miracle" rice may produce 80 percent grain (harvest ratio 4 to 1). The catch is that the "miracle" rice plant has no energy left for self-protection and requires a large amount of expensive auxiliary energy to nourish it and keep the bugs off—subsidies that small farmers and small nations often cannot afford. Also, the required chemical subsidies contribute to pollution. Accordingly, the Green Revolution is turning out to be a mixed blessing.

On the reference shelves of libraries, even small ones, you will find the United Nations' *FAO Production Yearbook*, which is issued annually and contains data on world food production. A sample of data from a recent yearbook is shown in Table 2. Annual yields of four crops that differ in protein content are shown at three levels: (1) the country with highest yield, (2) the country with lowest yield, and (3) the world average. High

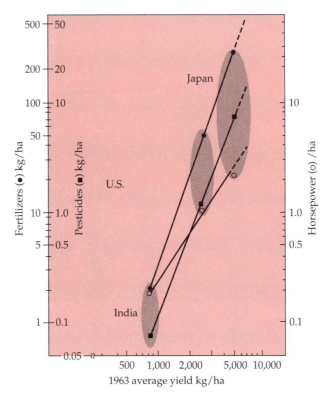

FIGURE 8
The relationship of energy and chemical subsidies to crop yields. (After
E. P. Odum 1983.)

production is found chiefly in European and North American countries
and Japan. About 30 percent of the world's people live in such developed
countries; 70 percent live in countries with food production one-third to
one-fourth less than that of the top countries. As a result, the world average is much closer to the bottom than to the top of the food production
range. From about 1980 to 1990, total food production in the undeveloped
world increased, but barely enough to keep up with population growth.
Furthermore, the increase came as much from increasing the amount of
land under cultivation as from increasing yield per unit of area. In contrast, yield increased greatly in Europe and North America while the
amount of land cultivated decreased—more food now comes from less
land. Yields of basic food crops have leveled off with very little increase

Thanks, but No Thanks

A great many people, including poorly informed political leaders, naively think that food production in Third World countries can be increased just by sending them seeds of high-yielding varieties of grains and a few "agricultural advisors." Cultivation of the "miracle" varieties requires expensive energy subsidies many undeveloped countries cannot afford. Until they can, it would be better to help such countries improve the labor-intensive agriculture already in place. In many cases these traditional or so-called "preindustrial" agricultural systems are quite sophisticated and energy-efficient.

since 1990, and the difference between rich and poor countries remains wide. Note also in Table 2 that the yield of a high-protein crop such as soybeans is considerably less than that of a low-protein crop such as sugarcane. As is often the case, quality reduces quantity (or vice versa).

It is protein rather than total calories that is lacking in the diets of many poor people worldwide. In Bangladesh, one of the world's poorest nations, protein consumption in 1960 was about 43 grams per day per

TABLE 2   Annual Yield of Edible Portion of Major Food Crops at Four Levels of Protein Content and Three Levels of Subsidy

Crops	Harvest Weight (Kilograms/Hectare)					
	Developed Country (with fuel-subsidized agriculture 1990)		Undeveloped Country (with little energy subsidy)		World Average	
					1990	1970
Sugar[a] (from sugarcane —less than 1% protein)	U.S.A.	7680	Pakistan	4150	6133	3000
Rice (10% protein)	Japan	6325	Bangladesh	2630	3550	2200
Wheat (12% protein)	Netherlands	7700	Argentina	1850	2550	1200
Soybeans (30% protein)	Canada	2650	India	1000	1900	1200

Figures are rounded-off averages from the *FAO Production Yearbook*, 1990. Yields of basic food crops have leveled off with very little increase since 1990, and the difference between rich and poor countries remains wide.

[a]Sugar is estimated as 10 percent of the harvest weight of the cane, as reported in the *Yearbook*

person—barely enough to sustain the body (compared with U.S. consumption of 100 grams or more). Thirty years later, in 1990, protein in the Bangladeshi diet had not increased—it was still 43 grams per day per person (1990 *FAO Production Yearbook*).

Meat is a major source of protein for humans. Animal production is at least one step down the energy chain (see Figure 3), so less meat than grain can be produced on a given area of land. Many people deeply concerned with widespread hunger in the world have suggested that we in developed countries should eat less meat so that more plant production can be available for human food. This might help some, provided the grain and soybeans we now feed to domestic animals could be somehow transported to the hungry, but we also have to remember that cows and fish can eat plant materials such as grass, algae, and detritus that is unsuitable as human food. Grass-fed beef and fish raised in ponds do not come at the expense of plant food for humans. It is now recognized that most people in affluent countries eat too much meat, especially fatty meat, so there are health reasons as well as economic reasons for eating less meat and more vegetables.

It is important to emphasize that good land on which to grow crops is in short supply. We have already noted that only a small fraction of the globe is naturally fertile. At best about 24 percent of the land area is arable, that is, suitable in terms of soils and climate for food production. Most of this is already in use for crops or pastures. Extending cultivation to marginal lands would be a big mistake, not only because of the high cost, but because of the damage to life-support ecosystems.

We need to be especially concerned about conversion of wet tropical forests to agriculture. The soils of many of these forests are easily leached of their few nutrients, as we shall see in Chapter 5, and therefore are not suitable for high-production, continuous row crop monoculture. Once depleted, these soils may be able to support only scrub or poor-quality pasture vegetation. This does not even consider the permanent loss of the rich fauna and flora of the tropical forests, a result which in itself would be tragic. In general, drier areas in the tropics (with irrigation where appropriate) are more suitable than are wet areas for sustainable agriculture. With millions of new hungry people appearing each year, it is going to be difficult to convince governments that the intact rain forest is more valuable in the long run than low-yielding cornfields and pastures. Only by setting aside large tracts now and by diverting agricultural development to drier lands can the biologically diverse tropical rain forests be saved. Several recent economic valuations (for example, Peters et al. 1989) indicate that the market value of nonwood products such as fruits, latex, or medicinals harvested on a sustained yield basis from rain forests can be much greater than that of a one-time timber harvest.

The Poor Get Poorer

When we compare the 1990 *FAO Production Yearbook* with the 1970 one, we find, sadly, that the gap between the food production of rich and poor countries has increased during the two decades by more than 2000 kilograms per hectare. The Green Revolution has helped many countries, such as India and China, feed their large populations, but overall the rich nations (and the wealthy large landholders in all nations) have benefited more than the poor nations (and small farmers everywhere) have (Shiva 1991). Exporting food from the "haves" to the "have-nots" helps, of course, but countries (and individuals) that need food the most can't pay for it and don't get it in today's free market—at least not until widespread starvation attracts the attention of the wealthy nations!

In 1967 and again in 1976, expert panels commissioned by the U.S. government issued lengthy reports (a total of nine volumes) on the "World Food Problem." These reports are "cautiously optimistic" that global famine can be avoided if the United States and other developed countries do more to improve the use and distribution of supplies and land, and to reduce the human birth rate. Norman E. Borlaug, a pioneer in the development of high-yielding varieties of cereal grains (and often called "the father of the Green Revolution") always includes in his writings and lectures a reminder that our success in increasing crop yields only buys a little time until the world population can be stabilized. Without a decrease in population growth, our efforts will have been futile.

Food for Domestic Animals

There is more to be considered than the food eaten by humans. There is also a huge population of domestic animals (e.g., cows, pigs, horses, sheep, and poultry) that consumes a large amount of the world's net primary production. In fact, domestic animals consume a lot more than humans do, since the standing crop biomass of livestock worldwide is about five times that of humans. The ratio of livestock to people (in biomass equivalent units) ranges from 43 to 1 in New Zealand, where sheep dominate the landscape, to 0.6 to 1 in Japan, where fish largely replaces terrestrial meat in the diet. The general ecology of a landscape, not to mention the culture and the economy, is much influenced by the kind of meat eaten. Thus, New Zealanders love (and economically depend on) their sheep, the Japanese worship their fish, and Americans have their love affair with the steak, the cowboy, and home on the range.

There are also pets, which in the United States consume a lot of fairly high-quality food—one can get an idea of how much by noting the number of shelves devoted to the display of pet food in the supermarket. In China there are few pets that are not also eaten.

Food from the Sea

At present, less than 5 percent of human food worldwide comes from aquatic ecosystems, and most of this is in animal form for reasons already noted (Emery and Iselin 1967). The percentages are higher in Japan, Southeast Asia, and North America, where there are productive estuaries, lakes, and streams. Fish harvests from the ocean have remained about the same for a number of years worldwide, with declines in many areas that have been overfished or polluted. Most fisheries biologists believe that it will not be possible to further increase the harvest of natural production. But artificial culture has possibilities. **Aquaculture** (plant and animal culture in water) is well developed in the Orient. Yields from managed ponds and enclosed sections of fertile estuaries equal or exceed beef production on land on a unit of area basis. Theoretically, fish production should be more efficient than cattle production since fish (and shellfish) are cold-blooded and do not have to dissipate energy to keep their bodies warm.

The problem is that there is not a lot of area suitable for intensive aquaculture, and fish and shellfish are especially vulnerable to disease, pollution, and cold weather. As in agriculture, the highest yields are obtained with subsidies. Adding mineral fertilizers to a pond can double or triple the yield of carp or catfish, while adding supplementary food such as grain or specially formulated high-protein fish chow can increase yields tenfold. While very little seafood consumed in the United States is cultured, about one-seventh of fish and shellfish consumed worldwide is aquacultured. One can expect to see an increase in this form of farming, but it is important to understand that the sea and the fresh waters are no bonanza waiting to be tapped to feed the hungry billions, as many people seem to think. To get high yields, expensive subsidies are needed, in water as on land.

Fuel and Fiber Production

Humanity draws heavily on primary production for agricultural products other than food, namely, fiber (e.g., cotton and paper) and fuel. For more than half the world's population, wood is the chief fuel used for cooking, heating, and light industry (Figure 9). In the poorest countries, wood is being burned much faster than it can be grown, so forests are turned into shrublands and then into deserts (when overgrazing stress is added). A 1975 report from the Worldwatch Institute (a private study group in Washington, D.C.) entitled *The Other Energy Crisis: Firewood* (Eckholm 1975) reports that in the African countries of Tanzania and Gambia, per capita fuelwood consumption is 1.5 tons per year and that 99 percent of the population uses wood as fuel. Since these countries do not have extensive forests, their future is grim.

In North America, and in other regions with large standing stocks of vegetation, there is considerable interest in using biomass from both forest and agricultural land as fuel. Among the options available are: (1) planting fast-growing trees harvested on short rotation (clear-cut and replant in ten years or less); (2) using limbs, stumps, roots, and other parts of the tree normally left in the forest after a timber harvest (the "complete tree harvest" concept); (3) reducing wood pulp demand by recycling paper and using pulp to generate electricity; (4) using agricultural plant

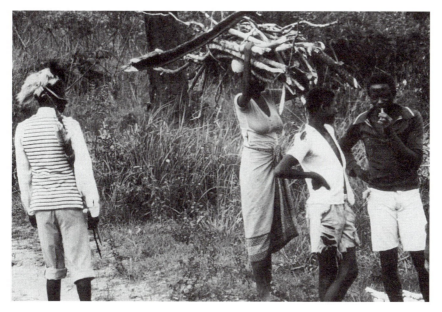

FIGURE 9
Wood gathering in Tanzania, where most of the human population uses wood as fuel. Note that it is the woman who bears the burden. (Photograph by P. R. Ehrlich, Stanford Univ./Biological Photo Service.)

and animal waste (manure) to produce methane gas or alcohol; and (5) growing crops such as sugarcane specifically for alcohol production for use in internal combustion engines.

Though all these options can certainly ease fuel shortages for the short term, all except the third have drawbacks: they could adversely affect the quality of soils, and they could increase the competition between food and fuel for arable land and worsen the already critical world food situation. According to a recent estimate, if motor fuel was obtained entirely from crops (alcohol and methane), at least one-fourth of the world's arable land would be required to satisfy the current global demand for motor fuel. According to Brown (1980), it takes 8 acres (3 hectares) to grow enough grain to run the average American automobile entirely on ethanol, while one-third of an acre can provide the amount of food eaten by an average Third World person.

When crops or trees are harvested, one third or so of the total biomass (e.g., stems, leaves, stumps, and roots) is left over. But, as soil scientist Hans Jenny points out, these materials are not "wastes;" they are extremely important in maintaining the fertility and water-holding capacity of soils. Jenny argues against indiscriminate conversion of biomass and organic by-products to fuel, because, in the long run, the "humus capital" is more valuable than fuel, especially since there are other sources for fuel, but not of humus (Jenny 1980). Ecologically speaking, the complete tree harvest concept seems an especially bad idea, since not only is all the organic matter removed, but the soil is ripped up as well. Brazil, which has to import most of its oil, is trying out option 5 above on a fairly large scale, and, from recent reports, with considerable success; many cars in that country are running on alcohol. It will be interesting to see how this works out in the long run, and whether it ultimately creates land use competition between food and fuel.

Energy Partitioning in Food Webs

Fortunately, human demands do not yet use up all of the biosphere's primary production. Vitousek et al. (1986) estimated that while only about 4 percent of terrestrial net primary production is used directly by humans and domestic animals as food, fiber, and wood fuel, some 34 percent more is co-opted by humanity in that it is part of nonedible production (such as lawns or the nonedible portions of crops) or is destroyed by human activity (such as the clearing of tropical forests or desertification). Estimates such as this are difficult to make and are subject to revision, but it would seem that at least 50 percent of terrestrial and most aquatic net production remains for "God's other creatures." It is vitally important that we not overlook the needs of the natural food webs that contribute to life support and global balance, which are already being stressed by our greed. This is why, in the preceding section, concern was expressed about human exploitation of the forest or the crop field. At least a third of primary production must be left for the ecosystem (it has to have food energy, too) if we expect to have any crops and forests in the future.

The general concept of the food chain and the food web has already been presented (see especially Figure 2 in Chapter 2 and Figures 2 and 4 in Chapter 3). Energy from the sun is transformed step by step by producers (plants), then primary consumers (plant eaters), then secondary consumers (animal eaters), and so on. We are all at least vaguely familiar with this sort of thing, since we eat the cow that eats the grass that fixes the sun's energy. Each step in the chain is called a **trophic level**. This is an energy level, not a species level, since any given species may utilize more than one level—humans, for example, are for the most part both plant and animal eaters. We have already documented how energy is lost at each transfer so that less is available at successive trophic levels. Thus, meat is more expensive than bread, and is much less common in the human diet when there are a great many mouths to feed.

The person who first introduced the concept of trophic levels to ecology was Raymond Lindeman, whose 1942 paper "The Trophodynamic Aspect of Ecology" is a classic. Unfortunately, Lindeman died at an early age; however, his ideas were further developed by G. Evelyn Hutchinson, who also pioneered the field of material cycling, as will be noted in the next chapter. (For a detailed account of Lindeman's contribution, see Cook 1977.)

In natural biotic communities with their great diversity of organisms, energy flow is not a simple linear process as in the grass–cow–human example, but a complex network of flows which we call the **food web**. A simplified model of the basic food web is shown in Figure 10. Plant production can become available to primary consumers as living material or as dead material (detritus); accordingly, we can designate consumption of the former as the **grazing food chain** and of the latter as the **detritus food chain**. Often a considerable part of primary production is in liquid form (dissolved organic matter, or DOM) that is exuded or extracted from living plant cells, flowers, and vascular systems. This pathway supports nitrogen-feeding microorganisms and nectar-feeding insects and hummingbirds, among others. The microbe-based food chain that develops from this energy source is called the **DOM–microbial food chain**. This pathway is especially important in the sea (Pomeroy 1974) and in the rhizosphere (root–soil zone). Cross-feeding between these food chains creates the food web.

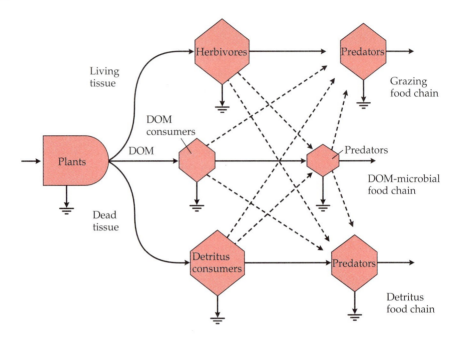

FIGURE 10
Generalized model of the food webs based on the three basic plant resources: living tissue, DOM (dissolved organic matter), and dead tissue (particulate detritus). The portion of net primary production energy that flows down these three primary pathways varies widely in different kinds of ecosystems and under different situations.

Most ecosystems (such as forests, oceans, and marshes) operate as detrital systems—less than 10 percent, often less than 5 percent, of plant production is grazed. This delayed consumption is important, since it allows a complex biomass to build up, increasing the storage and buffering capacity of the ecosystem. After all, a forest could never develop if young trees were grazed down as fast as they appeared. In some ecosystems—such as a pasture or grassland well stocked with large mammalian herbivores, or a pond where zooplankton are feeding heavily on phytoplankton—as much as 50 percent of net primary production may be grazed. But even in these situations, such a high grazing rate does not usually continue over an entire annual cycle.

Plants resist being eaten by producing antiherbivore chemicals such as tannins, alkaloids, or phenols, or by including in their tissues a lot of cellulose and woody material not digestible by most animals. Some plants, such as the grasses, are specially adapted to being grazed (and mowed) since their ground-level buds can send up new leaves almost as fast as they are clipped off, but there is a limit to how fast or for how long, of course. In any event, a great deal of plant material is not consumed until it dies and becomes particulate and dissolved organic matter (POM and DOM).

The DOM that leaks out of living plants is consumed largely by bacteria and fungi, which in turn are eaten by protozoa, microarthropods, nematodes, and other small animals that become food for larger ones. This kind of production is hard to measure, so there are few studies on the extent of its involvement in the food webs of different kinds of ecosystems. Included in the DOM-microbial food chain are the nitrogen fixers and mycorrhizal fungi, which may remove from 5 to 15 percent of a plant's photosynthate directly from its vascular system (Odum and Biever 1984; Paul and Kucey 1981). Recent studies have indicated that as much as 30–50 percent of primary production in the ocean is exuded as DOM from tiny floating plants and processed through a microbial food chain. In this habitat the DOM–microbial food chain may be more important than the other two (see Pomeroy 1974).

Grazing is a process that most everyone recognizes when cows or grasshoppers are involved. Detritus consumption, in contrast, is not very visible, and most people would not recognize a detritus consumer if they saw one, because many of these organisms are microscopic, or so small as not to be noticed. Actually, detritus consumption is a team process in which bacteria and fungi interact with small animals such as protozoans, nematodes, and mites, and in water, small crustaceans and insect larvae. The animals break up dead plant and animal material into bits and dissolved material, which are more readily available as food for bacteria and fungi than are large chunks. The animals also eat the microbes, which, strange as it may seem, stimulates these populations to grow faster and work harder. All these decomposers, of course, provide food for the higher trophic levels.

The partnership between microorganisms that can digest resistant plant residues (such as cellulose and wood) and animals that generally cannot is especially well developed in the ruminants (such as cattle, deer, and antelope) and termites. Ruminants have a special stomach, the **rumen**, where symbiotic microbes convert cellulose to sugars that the animal can use. Termites culture within their intestines specialized microbes (especially flagellates) that digest the wood and dead grass the termite eats, thereby providing nourishment for both partners. In some tropical grasslands termites consume more grass than the antelope and other grazers do, and their conspicuous mounds dominate the landscape (Figure 11). The next time you find termites in a decaying log, or in your house, you will know that you have before you not just one species, but one of the world's best detritus-consuming systems made up of cooperating species. And as is so often the case, what is a terrible pest in a human house is a worthwhile member of nature's house, where dead wood needs to be decomposed, not preserved.

The flowering plants that dominate terrestrial vegetation produce special products in connection with reproduction, namely, nectar, pollen, seeds, and fruit. These are nutritious foods that are fed upon by a diversity of specialists: the nectarivores, granivores, and frugivores. Nectar is produced by plants in order to attract pollinating insects and other animals. The nectarivorous pathway is especially prominent in tropical forests, where most species of plants are animal-pollinated, in contrast with temperate forests, where most species are wind-pollinated. Fruit attracts animals that aid in the dispersal of seeds. Both are energy-expensive productions, but necessary if the plant is to be propagated. These are the "wages" the plant pays for the survival of the next generation. These "wages" support an active "animal industry" that is an important part of the food web. Humans tap into a bit of this energy flow by cultivating honeybees.

FIGURE 11
Large termite mounds dominate these landscapes in Australia. (Photographs courtesy of John Alcock.)

When we consider f   e population and community levels, eating a   ne-way process in the
sense that only the eater   is killed and eaten by
a puma certainly does no   ory act, but the deer
population may be better   are taken by predators. The survivors will hav   nd the herd will be
less likely to exceed the re   derate grazing of
plants by herbivores is ofte   community as a
whole. Diversity tends to be ir   uces the numbers
of a dominant species. Grazin   ulate new plant
growth; in one study, a growt   as found in the
saliva of grasshoppers which, v   s as it is being
chewed on, stimulates the roots   des. The vast
herds of antelope on the East Afri   n to facilitate
the production of grass; net prima   ith the grazers than it would be without them   catch is that
the herds must move about over larg   ng. Fencing
in the game animals cancels out this

When a "downstream" organism in   ergy flow has a positive effect
of some kind on its "upstream" food supply, as in the examples just cited,
we have feedback, in this case positive feedback (recall Figure 2 in Chapter
2), or what we might even call **reward feedback.** Survival over the long run
is enhanced if the consumer can promote as well as utilize its food supply
(just as humans both consume and promote the welfare of their domestic
animals). The more food webs are studied, the more partnerships and
mutually beneficial relationships between producers and consumers and
between different levels of consumers are discovered (Lewin 1987; Dyer et
al. 1993, 1995). Contrary to what many people think, nature is not all "dog-
eat-dog." Competition and predation have their place, but survival often
depends on cooperation, as we shall see in Chapter 6.

Food chains vary in length for reasons that are not at all clear. Briand
and Cohen (1987) examined 113 food chains from the literature, and
found that length was independent of primary production, but that three-
dimensional or "thick" communities such as forests or ocean water
columns had longer chains than two-dimensional or "thin" communities
such as grasslands or intertidal seashores.

Energy Partitioning in the Individual

A model for energy partitioning in the individual or the species population
is shown in Figure 12. The shaded box, B, represents the living structure or
biomass. I represents the energy input: light in the case of autotrophs, food
in the case of heterotrophs. The usable part of the input is assimilated (A)

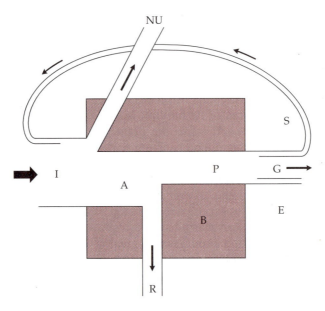

FIGURE 12
Partitioning of energy in the individual or population. I, input or ingested energy; NU, not used; A, assimilated energy; P, production; R, respiration; B, biomass; G, growth; S, stored energy; E, excreted energy.

and the unusable part ejected (NU). How much is assimilated depends on the quality of the energy source: as much as 90 percent if the food quality is high (sugar, for example) or as little as 5 percent if low (dead leaves, for example). A sizable portion of assimilated energy must always be respired to provide **maintenance** or **existence** energy to keep the body functioning and repaired; this is R in the diagram. What's left can be used for growth and reproduction, or it can be stored for future use (as fat, for example). This component is designated as production (P) in the diagram. Some small part may be lost in excretions (E).

How energy is partitioned between P and R is of vital importance to the individual and the species. Large organisms require more maintenance energy than small ones since they have more biomass to maintain. The relationship between rate of metabolism and size is not linear but curvilinear (more directly related to surface area than to weight). The warm-blooded animals, the birds and the mammals, respire more than the cold-blooded animals, which do not use energy to keep their body temperature up when it gets cold. Predators generally must expend a larger percentage of assimilated energy in respiration than herbivores, since a lot of energy has to be expended in finding and overcoming prey.

If it has a choice, a predator will seek larger, more nutritious prey items that require less expenditure of energy to consume; this leaves more energy available for reproduction or storage. Through natural selection, organisms achieve as favorable a benefit–cost ratio of energy use as possible. Ecologists spend a lot of time studying how optimization of energy use is accomplished in different kinds of organisms and under different conditions.

Allocation of energy to reproduction is, of course, vital to all organisms that are not bred by humans. Species adapted to unstable, recently denuded, or underpopulated areas generally allocate a large portion of their energy to reproduction, while species adapted to stable or fully populated communities allocate little energy to reproduction (most energy must go to survival of the individual under stressful conditions). A study of goldenrods, shown in Figure 13, illustrates this pattern. Species 1, found in open fields on disturbed soil, allocates 45 percent of its energy to producing flowers and seeds. At the other end of the spectrum, species 6, found in forests, puts less than 5 percent of its energy into reproduction.

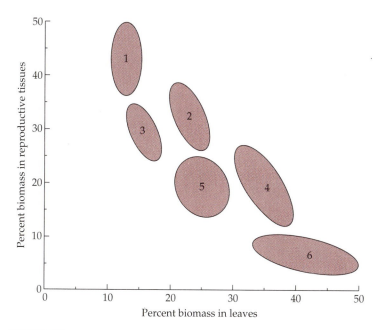

FIGURE 13
Partitioning of biomass between reproductive structures (flowers, seeds) and leaves in six species of goldenrod in habitat ranging from open fields (1) to forests (6). (After Abrahamson and Gadgil 1973.)

Its survival depends more on developing a large leaf area able to capture scarce sunlight—this is why you can't grow conspicuous ground-level flowers in the shade of trees. There are some interesting parallels here between natural and human populations that are considered in Chapter 6.

Energy-Based Classifications of Ecosystems

In Chapter 1, when the landscape was divided into three environments, namely, developed, cultivated, and natural, it was noted that from the standpoint of energy use we could categorize these as fuel-powered, subsidized solar-powered, and basic solar-powered, respectively. Now that we have reviewed energy principles, we can consider this sort of energy-based classification in more detail.

In Table 3, the **energy density** or **power level** (a concept also introduced in Chapter 1) for each of four ecosystem types is estimated in terms of kcal per square meter consumed annually. The basic solar-powered ecosystems have been divided into "unsubsidized" and "naturally subsidized" (recognizing that there can be no sharp dividing line between the two).

Ecosystems that run entirely or mostly on solar energy, such as the open ocean or upland forests, are the **unsubsidized solar-powered ecosystems**, category 1 in Table 3. Although low-powered (with an annual flow

Table 3 Ecosystems Classified According to Source and Level of Energy	
Ecosystem	**Annual Energy Flow [Power level] (kcal/m2)**
1. Unsubsidized natural solar-powered	1000-10,000 [2000]
2. Naturally subsidized solar-powered	10,000-40,000 [20,000]
3. Human-subsidized solar-powered	10,000-40,000 [20,000]
4. Fuel-powered urban-industrial	100,000-3,000,000 [2,000,000]

Numbers in brackets are estimated round-figure order-of-magnitude averages. The earth's ecosystems have yet to be inventoried in sufficient depth to calculate averages with any degree of confidence.

of about 2000 kcal per square meter), they cover a large part of the earth's surface and constitute a major part of our life-support environment. Organisms that populate these systems have remarkable adaptations for living on and efficiently using scarce energy and other resources.

Less common, but with an energy density about ten times higher, are the **naturally subsidized** and the **human-subsidized solar-powered ecosystems** (categories 2 and 3 in Table 3). Naturally subsidized ecosystems, such as tidal estuaries and some rain forests, are the naturally productive systems of nature that not only have high life-support capacity but also produce excess organic matter that may be exported to other systems or stored. Human-subsidized ecosystems are those supported by auxiliary fuel or other energy supplied by humans in order to produce food and fiber. Together these ecosystems provide most of our food, along with other valuable services. Looking down from an airplane or satellite, subsidized ecosystems appear bright green (due to an abundance of chlorophyll, indicating high rates of primary production). In the infrared photographs that are now routinely taken by orbiting Landsat spacecraft, they appear bright pink.

The rarest, but by far the most powerful, ecosystem types are the **fuel-powered urban–industrial ecosystems**. These are humanity's crowning achievements—the great cities and their industrial and suburban sprawl, which together are called **metropolitan districts**. Their energy density is several orders of magnitude (orders of 10) greater than that of solar-powered systems. The energy that flows annually through industrialized cities such as New York, London, and Tokyo is measured in millions rather than thousands of kcal per square meter. Highly concentrated fuel replaces rather than merely supplements the sun's energy, which instead becomes a costly nuisance by heating up the concrete and contributing to the production of smog. When we consider the high work capacity of fuel (as compared with solar energy), the difference in embodied energy between natural and urban–industrial systems is even greater than indicated by counting Calories. Industrialized cities are literally "hot spots," clumped together in certain regions such as the eastern and north central United States, Europe, and the central island of Japan, and widely scattered elsewhere as islands in a matrix of low-powered environments. Big industrial areas are, in fact, so energetic that cities such as New York and Washington, D.C., have a climate distinctly different from the countryside's. They are warmer, with more fog and drizzle and less sunlight due to the dust and smoke.

For the continental United States landscape as a whole, the average energy density directly related to human use of fuels is about 2000 kcal per square meter, about the same as for the average solar-powered natural

Killing the Golden Goose

Like that of producers and consumers in the food web, the city–country relationship is a mutually beneficial one, since the city gets its life-support goods and services from the country, which in turn benefits from the wealth and culture generated by the city. Unfortunately, political leaders very often do not seem to understand urban–rural interdependence; we constantly see country pitted against city in the political arena. It reminds one of children squabbling over toys, which is to say that much current political behavior is immature, short-sighted, and primitive, especially when we consider how mature and sophisticated our technology is. Let us hope that our political systems reach maturity before we kill all the country geese that lay the golden eggs for the city.

community (but remember that fuel and solar energy differ greatly in quality and, therefore, impact). The world average fuel energy density is only about 100 kcal per square meter (Smil 1984). As emphasized in Chapter 1, the fuel-powered city is so energetic that large areas of low-powered natural and agricultural environment are required to support it.

Energy Futures

At the First International Conference for the Peaceful Uses of Atomic Energy, held in Geneva, Switzerland, in 1955, the chairman of the conference, the late Homi J. Bhabha of India, described three ages of humankind: the age of muscle power (when animal, slave, and servant labor built great civilizations), the age of fossil fuel (when fossil-fueled "servant machines" freed the slaves), and the atomic age. Bhabha spoke eloquently of his belief that because of the universal availability of the atom, the coming atomic age would close the gap between rich and poor nations. I was a delegate at that meeting, and I was almost, but not quite, carried away by the enthusiasm for the coming utopia.

Thirty years later, the dream of cheap and abundant energy from the atom has yet to be realized, because tapping the enormous potential of atomic energy has proved to have a far greater "disorder potential" than anticipated in 1955, and the gap between rich and poor nations has grown progressively worse. Carroll Wilson, the first general manager of the U.S.

Atomic Energy Commission, expressed it this way in a 1979 article entitled "Nuclear Energy: What Went Wrong?": "No one appeared to understand (that) if the whole system does not hang together coherently, none of it might be acceptable." Ecologists who thought holistically did understand, but they were not able to articulate or document their concerns at that time.

It now appears that nuclear energy based on uranium fission is a flawed technology because there are no foreseeable solutions to the problems of the disposal of low- and high-level radioactive wastes (fission products), the high cost of fuel enrichment and plant construction, and the high risk of accidents such as those at Three Mile Island in the United States and Chernobyl in Ukraine. Until new, less disorderly ways to obtain energy from nuclear sources are devised, the coming of the atomic age has been postponed. Instead of peaceful uses of the atom, we now have mostly military uses and the terrible threat of nuclear war. In the meantime, the world seeks other sources of energy and more efficient (less wasteful) uses of the remaining fossil fuels to prolong their availability for as long as possible. One promising source of renewable energy involves tapping the temperature differential between warm surface water and deep cold water in tropical oceans to drive Rankin-cycle engines to generate electricity. This technology is known as Ocean Thermal Energy Conversion (OTEC) (see Avery and Wu 1994.).

The Net Energy Concept

What few people seem to realize is that it takes energy to produce energy, since some of the energy produced by any given conversion system has to be fed back to maintain the conversion system, as shown in Figure 14. To produce **net energy**, the yield (A) must be greater than the energy cost of sustaining the conversion system (B). For a power plant to be really worthwhile, the yield (net energy) should be at least two times, preferably four times, the energy cost (or "energy penalty," as engineers call it). For example, if 10 units of fuel are required to extract 12 units of oil by deep drilling under the ocean floor, this source is not very promising. Current uranium-fission power plants are so costly to build and maintain that net energy is marginal; various governmental subsidies (to pay for waste containment, for example) are often necessary to keep atomic power plants going. Experiments with nuclear fusion have so far failed to produce any usable net energy (energy cost greater than yield). The extremely high temperatures required for fusion to occur make it extremely difficult, perhaps impossible, to tame the H-bomb.

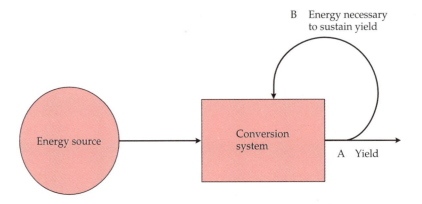

A − B = Net energy

FIGURE 14
The concept of net energy. The yield must be greater than the energy
necessary to sustain the system for the conversion to achieve a positive
net energy value.

   Despite the difficulties, atomic energy in some form should have a
future if, as already noted, we can find less disorderly ways to tap its
potential. In the meantime, *matching the quality of source with use* becomes
especially important. For example, if low-concentration solar energy
could be used for heating homes and buildings (low-quality work), then
highly concentrated energy such as oil and electricity could be saved for
the high-quality work of driving machines. When you think about it, it is
a terrible waste to burn oil just to heat a house, because a very high-quality source is being used for the lowest quality work. California is providing leadership in this area, since most new homes in that state are
designed to obtain at least some space and water heating from the sun.
For the immediate future, at least, it seems that humans may need to use
several sources of energy rather than depending on one dominant source
as we have for the past century.
   A recent report by the International Energy Agency states that conservation efforts have increased the efficiency with which energy is used in
Western industrialized nations by about 20 percent (a saving of energy
equivalent to 880 million tons of oil a year; see the 1987 news report in
Science 135: 23–25). Continued efforts to increase energy conservation can
save even more. While we cannot predict the future with any certainty, we
can recognize and avoid what would be undesirable futures, such as will
result if we waste highly concentrated energy, which will be in increas-

Taking the "Soft Path"

I don't know anyone more patient than Amory Lovins and his wife Hunter when it comes to energy futures. For several decades now they have been documenting, writing, lecturing, and testifying for what they call the "soft path," which involves energy conservation and seeking solar alternatives to fossil fuels (Lovins 1977, 1991; Lovins and Lovins 1986). Until recently their message has fallen on deaf ears, but now even the utility companies are becoming convinced that energy-saving measures are cheaper and better for the environment than constantly seeking to get more energy. For example, doubling the efficiency of electrical use and insulating houses and buildings will greatly reduce the need to build ever more power plants.

ingly short supply, and if industrialized countries do not adopt efficiency measures to reduce per capita energy consumption.

Energy and Money

Money is directly related to energy, since it takes energy to make money. Money is a counterflow to energy in that money flows out of cities and farms to pay for the energy and materials that flow in. The trouble is that money tracks human-made goods and services but not the equally important natural goods and services, as shown in Figure 15. At the ecosystem level, as shown in Figure 15A, money enters the picture only when a natural resource is converted into marketable goods and services, leaving unpriced (and therefore not appreciated) all the work of the natural system that sustains this resource. In the example shown, only the harvest and seafood processing part of the production chain is valued in terms of money; all the energy and work performed by the estuary to sustain the crop and to provide other valuable services such as recycling of air and water are entirely external to the monetary system. Accordingly, the estuary is worth a great deal more to society as a whole than is indicated by the economic value of its products. It is worth a lot even if no products are harvested.

At the global level, as shown in Figure 15B, money flows ($) accompany energy flows from human-made and domesticated ecosystems, but not from natural systems. In other words, we pay for goods and services from urban–industrial and agricultural ecosystems but not for the goods

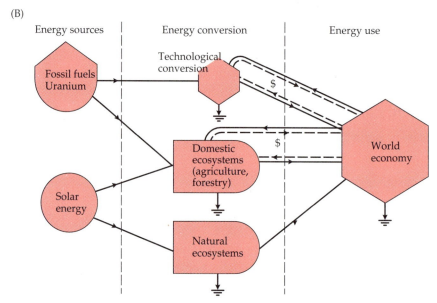

FIGURE 15

Energy and money. (A) Value of an estuary. In conventional economics, money is not involved until fish are caught; the work of the estuary that produces fish is given no value. The total value of the estuary in terms of useful work for humans is at least ten times the value of the harvested products. (Solid arrows represent energy flows; broken arrows represent money flows.) (See Gosselink et al. 1974.) (B) The energy support system for humans. Money flows ($) accompany energy flows from human-made and domesticated ecosystems, but not from natural ecosystems. (Diagram by H. T. Odum.)

Do We Need A New Currency?

Because money and energy are related, using energy as a basis for evaluating and allocating goods and services of all kinds is a logical approach to dealing with the fact that money alone is an incomplete measure of wealth (H. T. Odum 1973, 1996; Hall et al. 1986.) For example, to evaluate the total worth of an estuary, the total energy flow in terms of eMergy (which represents all the work of the ecosystem) would be determined, then this value would be converted to monetary units on the basis of the ratio between energy and money in the production of market goods. If it takes 10^4 Calories to produce one dollar in the local or national economy (ratio of per capita energy consumption to per capita income, for example), then an energy flow of $1000 \times 10^4 = 10^7$ Calories per acre per year would indicate a value of $1000 for all the goods and services produced during that year. Prorated over a period of years, the acre of estuary would have a value equivalent to tens of thousands of dollars in the market economy. Gosselink et al. (1974) used this approach to estimate the total value of tidal estuaries at $20,000–$50,000 per acre.

Another approach to correcting market failure involves increasing the citizen's willingness to pay for services to sustain the life-supporting biosphere. For example, it seems likely that most people would be willing to pay 10–20 percent more for electricity in exchange for cleaner air. Air pollution and acid rain would be much reduced if power plants removed contaminants in coal and other fuel before it was burned. Such "clean fuel technology" is already well developed and is in use in experimental coal-burning plants in California (Spencer et al. 1989). A logical consequence of higher fuel prices should be the construction of energy-efficient houses and buildings, which would further reduce the stress on our atmosphere as well as our utility bills.

and services from natural ecosystems. Accordingly, there is a market failure when it comes to the latter values.

Kenneth Boulding is one of the few economists who has been elected to the prestigious National Academy of Sciences. For three decades he has been arguing for the development of a more holistic economics that would close the gap between market (priced), nonmarket (unpriced) values. His numerous books and articles have provocative titles, such as *A*

Reconstruction of Economics (1965), "The Economics of the Coming Spaceship Earth" (1966) and *Ecodynamics* (1981). His writings were widely cited but had little effect on economic practices of the 1960s and 70s. However, as reviewed in the Epilogue, a good dialogue between economists and ecologists has been established in recent years. In fact, an economist and an ecologist have gotten together to edit a journal, *Ecological Economics*.

Money is one of our most important inventions, and it is now the basis for decision-making at most levels of society. But we must remember that our money system does not now take into account all the real costs of living, and we must take care that money is not allowed to be the only factor in the decisions we make. When it comes to the quality of human life, money and the consumption of human-made market products (on which current economies are based) are not the only considerations. Literacy, aesthetics (appreciation of art, music, and nature), love, and health deserve higher priorities in our never-ending pursuit of happiness.

Suggested Readings

*Abrahamson, W. G., and M. Gadgil. 1973. Growth form and reproductive effort in goldenrods (*Solidago*, Compositae). *Am. Nat.* 107: 651–661.

Adey, W. H. 1987. Food production in low-nutrient seas. *BioScience* 37: 340–348. (Shellfish and crabs cultured on platforms—artificial reefs—suspended in the lighted zone.)

Altieri, M. A., D. K. Letourneau, and J. R. Davis. 1983. Developing sustainable agroecosystems. *BioScience* 33: 45–49.

*Avery, W. H., and C. Wu. 1994. *Renewable Energy from the Ocean: A Guide to OTEC*. Oxford University Press, Oxford.

Black, C. C. 1971. Ecological implications of dividing plants into groups with distinct photosynthetic capacities. *Adv. Ecol. Res.* 7: 87–114.

*Bowman, K. P. 1988. Global trends in total ozone. *Science* 239: 48–50.

*Boulding, K. E. 1965. *A Reconstruction of Economics*. Science Editions, New York.

*Boulding, K. E. 1966. The economics of the coming spaceship earth. In *Environmental Quality in a Growing Economy*. Johns Hopkins University Press, Baltimore.

*Boulding, K. E. 1981. *Ecodynamics*. Sage Publications. Beverly Hills, CA.

*Briand, F., and J. E. Cohen. 1987. Environmental correlates of food chain length. *Science* 238: 956–960.

*Brown, L. R. 1980. *Food or Fuel: New Competition for the World's Cropland*. Worldwatch Paper no. 35. Worldwatch Institute, Washington, D.C.

*Cohn, J. 1987. Chlorofluorocarbons and the ozone layer. *BioScience* 37: 647–650.

*Commoner, Barry. 1976. *The Poverty of Power: Energy and Economic Crisis*. Bantam Books, New York.

Cook, E. 1971. The flow of energy in an industrial society. *Sci. Am.* 224, 225(3): 135–144.

*Cook, R. E. 1977. Raymond Lindeman and the tropho-dynamic concept in ecology. *Science* 198: 22–26.

*Dyer, M. I., A. M. Moon, M. R. Brown, and D. A. Crossley. 1995. Grasshopper crop and mid-gut extract effect on plants: an example of reward feedback. *Proc. Natl. Acad. Sci.* 92: 5475–5478.

Dyer, M. I., C. L. Turner, and Tr. R. Seastedt. 1993. Herbivory and its consequences.. *Ecol. Appl.* 3: 10–16.

*Eckholm, E. P. 1975. *The Other Energy Crisis: Firewood*. Worldwatch Paper no. 1. Worldwatch Institute, Washington, D.C.

*Emery, K. O., and C. O.'D. Iselin. 1967. Human food from ocean and land. *Science* 157: 1279–1281.

*Food and Agricultural Organization of the United Nations. 1990. *FAO Production Yearbook*, vol. 44. Food and Agricultural Organization, Paris.

Gates, D. M. 1985. *Energy and Ecology*. Sinauer Associates, Sunderland, MA.

*Gosselink, J. G., E. P. Odum, and R. M. Pope. 1974. *The Value of the Tidal Marsh*. LSU-SG-74-03. Center for Wetland Resources, Louisiana State University, Baton Rouge.

*Hall, C. A. S., C. J. Cleveland, and R. Kaufmann. 1986. *Energy and Resource Quality: The Ecology of the Economic Process*. Environmental Science and Technology. Wiley, New York.

*Hulbert, M. K. 1971. The energy resources of the earth. *Sci. Am.* 224, 225(3): 60–70.

*Jenny, H. 1980. Alcohol or humus? *Science* 209: 444.

*Kerr, R. A. 1996. Ozone-destroying chlorine tops out. *Science* 271: 32.

Lewin, R. 1987. On the benefits of being eaten. *Science* 236: 519–520. (Review of recent work on herbivore–plant and predator–prey positive feedback.)

Lewin, R. 1989. Sources and sinks complicate ecology. *Science* 243: 477–478.

Lieth, H., and R. H. Whittaker. 1975. *Primary Productivity of the Biosphere*. Ecological Studies, vol. 14. Springer-Verlag, New York.

Lieth, H., and R. H. Whittaker. 1975. *Productivity of World Ecosystems*. National Academy of Sciences, Washington, D.C.

*Lindeman, R. L. 1942. The trophic–dynamic aspect of ecology. *Ecology* 23: 399–418.

Lotka, A. J. 1922. A contribution to the energetics of evolution. *Proc. Natl. Acad. Sci. USA* 8: 140–155.

*Lotka, A. J. 1925. *Elements of Physical Biology*. Williams and Wilkins, Baltimore.

*Lovins, A. B. 1977. *Soft Energy Paths*. Ballinger, Cambridge, MA.

*Lovins, A. B. 1991. Energy conservation. *Science* 251: 1296–1298. (See also *Science* 252: 763.)

*Lovins, A. B., and H. Lovins. 1986. *Energy Unbound: Your Invitation to Energy Abundance*. Sierra Club Books, San Francisco.

*McNaughton, S. J. 1976. Serengeti migratory wildebeest: Facilitation of energy flow by grazing. *Science* 191: 92–94.

*Odum, E. P., and L. J. Biever. 1984. Resource quality, mutualism, and energy partitioning in food chains. *Am. Nat.* 124: 360–376.

*Odum, H. T. 1971. *Environment, Power, and Society*. Wiley-Interscience, New York.

*Odum, H. T. 1973. Energy, ecology, and economics. *Ambio* 2(6): 220–227.

*Odum, H. T. 1996. *Environmental Accounting: EMergy and Environmental Decision Making*. John Wiley & Sons, New York.

*Odum, H. T., and E. C. Odum. 1981. *Energy Basis for Man and Nature*, 2nd ed. McGraw-Hill, New York. (The maximum power principle is explained on pp. 32–34.)

*Paul, E. A., and R. M. N. Kucey. 1981. Carbon flow in plant microbial associations. *Science* 213: 473–474.

*Peters, C. M., A. H. Gentry, and R. O. Mendelsohn. 1989. Valuation of an Amazon rain forest. *Nature* 339: 655–656.

*Pomeroy, L. R. 1974. The ocean's food web, a changing paradigm. *BioScience* 24: 499–504.

*Pulliam, H. R. 1988. Sources, sinks, and population regulation. *Am. Nat.* 132: 652–661.

*Prigogine, I., G. Nicoles, and A. Babloyantz. 1972. Thermodynamics and evolution. *Physics Today* 25(11): 23–38; 25(12): 138–141.

Prigogine, I., and I. Stengers. 1984. *Order out of Chaos: Man's New Dialogue with Nature.* Bantam, New York.

Rowland, F. S. 1989. Chlorofluorocarbons and the depletion of stratospheric ozone. *Am. Sci.* 77: 36–45. (Review by one of the two persons who first suggested, in 1974, that chlorofluorocarbons are destroying the ozone layer.)

*Shiva, V. 1991. The green revolution in the Punjab. *Ecologist* 21(2): 57–60. (Negatives: increased vulnerability to pests; soil erosion; water shortages; displacement of small farmers. Beneficiaries: agro- and petrochemical companies; large landowners.)

*Smil, V. 1984. On energy and land. *Am. Sci.* 72: 15–21.

*Spencer, D. F., A. S. Alpert, and H. H. Gilman. 1986. Cool water: Demonstration of a clean and efficient new coal technology. *Science* 232: 609–612.

Starr, C. ed. 1971 *Energy and Power.* Special issue, *Sci. Am.* 224, 225(3).

Starr, C., M. F. Seari, and S. Alpert. 1992. Energy sources: A realistic outlook. *Science* 256: 981–987. ("Unabated historical energy trends would lead to an annual energy demand four times the present level by the middle of the next century, but extensive global conservation and energy-efficient systems might reduce this value by half.")

Sun, M. 1984. Pests prevail despite pesticides. *Science* 226: 1293. (Report on an international conference on the subject.)

*Vitousek, P. M., P. R. Ehrlich, A. H. Ehrlich, and P. A. Matson. 1986. Human appropriation of the products of photosynthesis. *BioScience* 36: 368–373.

*Whittaker, R. H., and G. E. Likens, eds. 1971. Primary production of the biosphere. *Hum. Biol.* 1: 301–369.

*Wilson, C. L. 1979. Nuclear energy: What went wrong? *Bull. Atom. Sci.* 35(6): 13–17.

*Indicates references cited in this chapter.

5

Material Cycles and Physical Conditions of Existence

The big-city television meteorologist is ecstatic when he or she can forecast a rainless day and apologizes profusely when there is rain in the forecast, especially if it is the weekend. Every day should be sunny, and a rainy day is viewed as a disaster because it makes life a bit uncomfortable or inconvenient. Yet if it never rained, there would be no life, no city. The urbanite's view of the weather illustrates our tendency to be short-sighted in regard to our environment. Farmers, of course, appreciate rain, but complain just as loudly when the weather doesn't suit the immediate needs of their crops. We cannot live without weather—and it is a universal subject of conversation. The nineteenth-century writer John Ruskin was on the mark when he said, "There is really no such thing as bad weather, only different kinds of good weather." Even storms provide environmental benefits, such as bringing in micronutrients from the sea and recharging groundwater, and their costs would be less severe if people did not insist on building on floodplains or too close to shores.

Water, a major component of our weather, is a vital life-supporting material that cycles back and forth between living organisms and the abiotic environment. As water is used, it evaporates from vegetation, lakes, and other surfaces, percolates through soil to become groundwater, and runs off in streams and rivers to the sea. No matter how water leaves the ecosystem, it must be eventually replaced by rain (or prehistoric rain stored as underground water) if commerce, agriculture, recreation, or any part of human life is to continue as before.

The Hydrological Cycle

The water cycle, or in technical terms the **hydrological cycle**, is shown in Figure 1 in its two phases: the uphill or upstream phase, driven by solar energy, and the downstream phase, which provides the goods and services that we and our environment require. More water evaporates from the sea than returns there as rainfall, and vice versa for the land. Thus, a

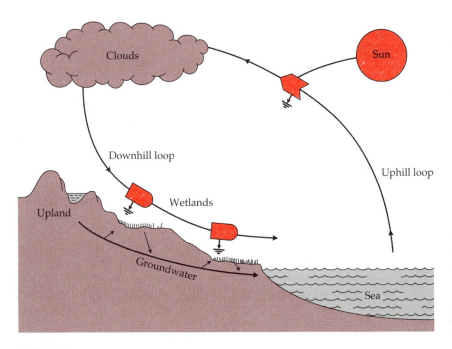

FIGURE 1
The energetics of the hydrological cycle as viewed in two loops: the uphill loop driven by solar energy, and the downhill loop that releases energy to lakes, rivers, and wetlands and performs useful work of direct benefit to humans (such as hydropower).

Clean Water for Free

As noted in Chapter 4, about one-third of the solar energy reaching the earth's surface is dissipated in driving the hydrological cycle. The natural recycling of water is not only a "free" service, but constitutes a vast use of solar energy that is almost unrecognized by the general public. It will be a sad day for our economy if we are forced to use expensive fuels to recycle and purify all drinking water. And this could very well happen if we fail to prevent contamination of our fresh water sources. A gallon of artificially desalted and filtered seawater delivered to your home would cost at least a dollar, as it now does on some remote islands with little rainfall. With nature doing most of the work, your water now costs less than a dollar for thousands of gallons!

---

considerable part of the rainfall that supports land ecosystems and most human food production comes from water evaporated from the sea. For example, an estimated 90 percent of the rain in the Mississippi River valley comes from the sea (the Gulf of Mexico, chiefly). Hydroelectric power is a direct benefit we obtain from the energy of downhill water flow.

Quantifying the hydrological cycle, that is, putting numbers on the flows, is a difficult task that is not yet complete. One estimate is that 20 percent of annual rainfall on land runs off to the sea, and 80 percent recharges the surface and groundwater reservoirs. Humans increase runoff and decrease infiltration into the soil and the water table by paving, ditching, draining swamps, compacting soils, and cutting down forests. Groundwater (i.e., the water you seek when you dig a well) is more plentiful than surface water in many regions and is being increasingly used by humans for irrigation, industry, and drinking water. The amount of groundwater in the entire continental United States is estimated to be four times the volume of the Great Lakes. Despite this large supply, the United States is threatened with water shortages due to the lack of groundwater recharge and contamination of the groundwater by toxic substances.

The Ogallala Aquifer Dilemma

The largest stores of groundwater are in **aquifers**: porous underground strata, often limestone, sand, or gravel, bounded by impervious rock or clay, that hold water like a giant pipe or elongated tank. In arid regions, many aquifers are not being recharged, or are recharged at such a low rate that the water in them must be considered a nonrenewable resource like fossil fuel (accordingly, we can speak of it as fossil water). In the western United States, about one-fourth of the withdrawals from aquifers are now considered to be overdrafts (exceeding recharge). An example is the Ogallala aquifer in the high plains of Texas, Kansas, Oklahoma, Nebraska, and eastern Colorado. Irrigated grain production in this region provides an important part of the export market that the United States counts on to balance payments for imported oil. Fossil water and fossil fuel (to pump the water) have combined to create a billion-dollar economy in the region. Unfortunately, the aquifer will, for all practical purposes, be "pumped out" by or shortly after the year 2000 (Opie 1993). The water will be gone before the fossil fuel, but the fuel becomes useless without the water. The region will then presumably experience severe economic depression and depopulation as land use returns to much less lucrative dryland farming. The dust storms of the 1930s could return when the water now used to keep the landscape green is gone (Figure 2).

Worldwide, about 70 percent of all water pumped out of aquifers is used to irrigate crops. As noted in Chapter 4, irrigation, the increased use of fertilizers, and improved, high-yielding crop varieties have all combined to increase crop yields in many parts of the world. The big question now is, How sustainable is this water use? How long can we continue to increase pumping in our efforts to feed ever more and more hungry

Alternatives to an Aquifer?

A possible way to continue irrigated farming if the Ogallala Aquifer were pumped out might be to build an aqueduct from the Mississippi River system. Such a venture would be expensive and would raise serious political and ethical questions. Should the nation's taxpayers be required to bail out a region where the people deliberately exhausted their chief natural resource? Or would it be better gradually to move the grain farming to a better-watered region before the water is all gone? And if so, who would organize and carry out a planned transition—the states, the federal government, or the agricultural industry?

FIGURE 2
The dust storms of the 1930s could return to the high plains of the United States when the groundwater now used to keep the landscape green runs out early in the next century. We know enough about soil conservation to prevent such a catastrophe, but short-term economic considerations make it difficult to plan for a transition. (Photograph courtesy of Soil Conservation Service.)

humans and domestic animals? For desert countries such as Libya and Saudi Arabia, the answer is, not for long (Gardner 1995).

Salinization

In addition to depleting the groundwater supply, crop irrigation can cause **salinization**, or the buildup of salts as water evaporates in the fields, especially in warm, dry regions. More irrigated cropland is currently being lost as a result of salinization than because of water shortages. A great deal of research on ways to reduce this threat is now in progress.

Speaking of salts, the use of salt (sodium chloride) on roads for ice and snow removal has increased to such an extent that roadside trees and other vegetation are being destroyed and underground water mains, telephone lines, and electric cables are being corroded. Seepage of salt into groundwater aquifers is beginning to be recognized as a health hazard for humans. More than 10 percent of all salt produced in the world annually is now applied each year to American highways in snowy states!

Biogeochemical Cycles

Ecologists term the more or less circular paths of the chemical elements passing back and forth between organisms and environment **biogeochemical cycles**. *Bio* refers to living organisms and *geo* to the rocks, soil, air, and water of the earth. Geochemistry is an important physical science, concerned with the chemical composition of the earth's crust and its oceans, rivers, and so on. Biogeochemistry is thus the study of the exchange (that is, the back-and-forth movement) of materials between the living and nonliving components of the biosphere. The term probably originated with the Russian scientist Vladimir Ivanovich Vernadsky (1863–1945), who is best known for his book *The Biosphere* (1926), but the field was pioneered by G. Evelyn Hutchinson (1944, 1950).

In Figure 3, a biogeochemical cycle is superimposed on a simplified energy flow diagram to show the interrelation of the two basic processes.

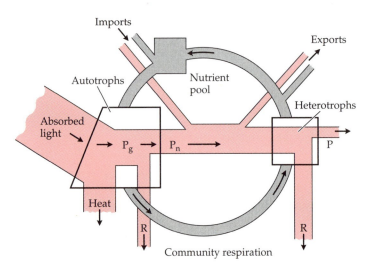

FIGURE 3
Relationships between the cycling of nutrients and the one-way flow of energy that drives biogeochemical cycles. P_g, gross primary production; P_n, net primary production; P, secondary production; R, respiration.

Remember that energy is required to drive the cycling of materials. Natural recycling is mostly driven by natural energy such as sunlight. For artificial recycling to have a net benefit, work energy has to be available at a cost not exceeding the value of the recycled product. When resources are abundant and supply far exceeds demand, artificial recycling is not appropriate, except for very valuable materials such as gold or platinum. But when supplies become limited, artificial recycling becomes feasible and desirable, as will be illustrated by paper recycling later in this chapter. Organisms in nature operate the same way: they tend to hoard and recycle vital elements, such as phosphorus, that are relatively scarce in terms of the need for them.

Like water, the two dozen or so vital nutrient elements (carbon, nitrogen, phosphorus, calcium, potassium, and others required by living organisms in varying degrees) are not homogeneously distributed or present in the same chemical form throughout an ecosystem. Rather, materials exist in **compartments** or **pools**, with varying rates of exchange between them. In general, it is practical to distinguish between a large, slow-moving nonbiological pool and a smaller, more active pool that is exchanging rapidly with organisms. For example, the soil of your garden contains phosphorus in an insoluble form that is not immediately available to the roots of your flowers or vegetables, but it also contains soluble phosphorus that can be absorbed and used by the plants during the growing season. Often the available pool is small, so one needs to add fertilizer to obtain high yields.

In Figure 3 the large reservoir is the box labeled "nutrient pool," and the rapidly cycling material is represented by the gray circle going from autotrophs to heterotrophs and back again. Sometimes the reservoir portion is called the unavailable pool and the cycling portion the available pool; such a designation is permissible provided it is clearly understood that these terms are relative. An atom in the reservoir pool is not necessarily permanently unavailable to organisms. Almost always there is a slow movement between unavailable and available pools.

Decomposition releases not only minerals but also organic by-products that may affect the availability of minerals to autotrophs. One way this occurs is by a process known as **chelation** (Fr. *chele*: "claw," referring to grasping) in which organic molecules "grasp" or form complexes with calcium, magnesium, iron, and other ions. Chelated minerals are more soluble and often less toxic than some of the inorganic salts of the same element, especially in the case of metals. For example, copper in industrial wastes is less toxic to marine organisms in inshore waters, where there is a lot of organic matter, than in offshore waters, where chelating substances are less abundant.

Two Basic Types of Cycles

From the standpoint of the biosphere as a whole, biogeochemical cycles fall into two groups: gaseous types with a large reservoir in the atmosphere, and sedimentary types with a reservoir in the soils and sediments of the earth's crust. The nitrogen (N) cycle and the phosphorus (P) cycle, illustrated in Figures 4 and 5, are good examples of these two types. Since nitrogen and phosphorus are important nutrients whose availability often limits productivity, it is important that we understand how these vital minerals behave. In general, nitrogen is more likely to be limiting to primary production in the sea, while phosphorus is often the limiting nutrient in fresh water. Both are often in short supply in terrestrial soils.

The Nitrogen Cycle

Nitrogen (N) is continually feeding into and out of the atmospheric reservoir and the rapidly recycling pool associated with organisms. Both biological and nonbiological mechanisms are involved in **denitrification**, which puts nitrogen into the air, and **nitrogen fixation**, the conversion of gaseous nitrogen, which is not usable directly by autotrophs, into ammonia, nitrite, and nitrate, which are usable. Specialized microbes play key roles in most of the steps in the nitrogen cycle, as shown in Figure 4. For example, only a few primitive bacteria (prokaryotes, including the blue-green algae, which are more properly called cyanobacteria) can fix nitrogen. Legumes and some other higher plants fix nitrogen only through the prokaryotic bacteria that live in special nodules on their roots. This is another example of the vital role played by microorganisms in maintaining our life-support systems, as was mentioned in the discussion of the Gaia hypothesis in Chapter 3. The feedback and exchange mechanisms, as shown in a simplified manner in Figure 4, make the nitrogen cycle and other cycles (such as carbon and water) effectively self-adjusting over large areas of the biosphere. Any increase in movement along one path may be quickly compensated by adjustments along other paths. Locally, however, nitrogen often becomes limiting to the biological system, either because regeneration (that is, movement from unavailable to available states) is too slow or because a net loss is occurring from the area.

The availability of nitrogen is of paramount importance to us and our fellow creatures because nitrogen is a necessary part of the basic units of all life: DNA (the genetic material), proteins, and amino acids. Yet getting nitrogen from the huge atmospheric reservoir that surrounds us to our cells involves a long, energy-consuming sequence of fixations and food chains. A lot of energy is required to break the triple bond of atmospheric

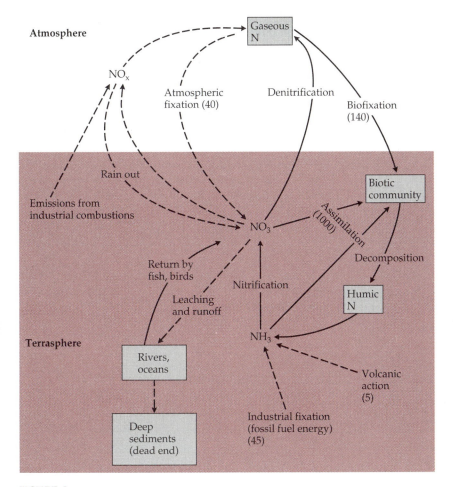

FIGURE 4

The nitrogen cycle. Nitrogen cycles in its several major forms between the large storage pool in the atmosphere and the smaller, but more active, pools in the soils and water of the earth itself (terrasphere). Solid arrows indicate the flows and exchanges mediated and controlled by organisms (especially microorganisms). Dashed lines show the flows that result chiefly from physical forces or are the result of human activities. Nitrogen oxides (NO_x) in the atmosphere can be in several forms, such as NO_2 or N_2O. It is these forms that contribute to acid rain, smog, and other forms of air pollution. At the present time it is not possible to put numbers on all the flows at the global level. The numbers shown in parentheses are there merely to show the relative importance of some exchanges (for example, biofixation by nitrogen-fixing microorganisms is more important than fixation by lightning or industrial fixation during the production of fertilizers).

nitrogen (N-N) so that nitrogen can react with water (H_2O) to produce ammonia (NH_3).

The Phosphorus Cycle

Most nutrients are more earthbound than nitrogen; their cycles are less self-adjusting and, consequently, more easily disrupted by humans. The phosphorus cycle (Figure 5) is a good example of a sedimentary cycle. Phosphorus is required for the energy transformations that distinguish living protoplasm from nonliving material, and it is relatively rare on the earth's surface compared with the biological demand for it. Organisms have devised many mechanisms for hoarding this element; hence, the concentration of phosphorus in a gram of biomass is usually many times that in a gram of surrounding environment (water or soil, for example).

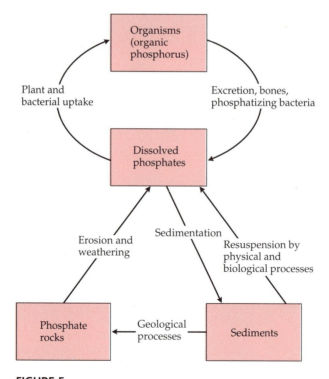

FIGURE 5
A simplified diagram of the phosphorus cycle, including slow fluxing to and from the sedimentary reservoirs.

As the cycling pool of phosphorus spins through organisms and local biogeochemical cycles, there is a tendency toward a slow downhill movement of reservoir phosphorus into the sea, following the pattern of erosion and sedimentation. In the long run, phosphorus is returned to the land by mountain building, the weathering of rocks, airborne dust or salt spray, and volcanic gases, which may move the mineral uphill. Wind-driven upwelling of deep ocean waters brings phosphorus and other nutrients from the unlighted depths to the photosynthetic zone (where it can be used by plants) and provides an important return mechanism in the sea. Fish-eating birds return many tons in the phosphorus-rich guano they deposit on coastal nesting grounds, where it is often harvested for use as fertilizer.

Humans have so increased the rate of soil erosion and runoff that the one-way movement of phosphorus into the large unavailable ocean pool has increased. For the present, agriculturists are not worried, since there are considerable reserves of phosphate rock that can be mined to replace some of the loss from cultivated land. This activity, however, creates severe local environmental problems, as visitors to the phosphorus strip mines southeast of Tampa, Florida, rapidly become aware.

As emphasized in our discussion of input management in Chapter 1, it is becoming increasingly necessary to improve the retention and recycling of phosphorus and other nutrients in agricultural and other human-ordered systems, not only to conserve supplies but to reduce nonpoint pollution (runoff into surface water and groundwater). Hopefully, we will never have to try to retrieve phosphorus from the deep sea, which will cost us energy and money and greatly increase the price of food.

The Sulfur Cycle

The sulfur (S) cycle, shown in Figure 6, illustrates many of the main features of material cycling.

1. There is a large reservoir in sediments and a smaller reservoir in the atmosphere.
2. The key role in the rapidly fluctuating pool (the center "wheel" in Figure 6) is played by organisms, principally by specialized microorganisms that function like a relay team, each carrying out a particular chemical transformation (the numbered steps in Figure 6).
3. A gaseous phase, hydrogen sulfide (H_2S), moves upward, which results in a microbial recovery of sulfur otherwise "lost" in the deep sediments.
4. The interaction of geochemical, meteorological, and biological processes and the interdependence of air, water, and soil maintain the cycle at the global level.

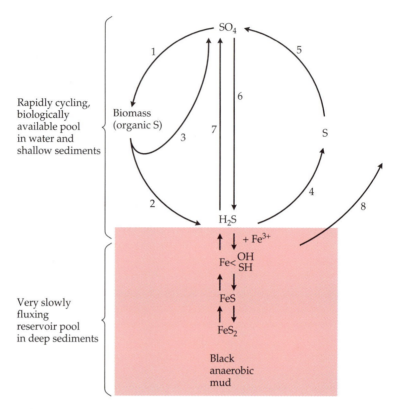

FIGURE 6

The sulfur cycle. Step 1 is primary production by plants. Organisms, most of them specialized microorganisms, carry out steps 2–7: 2, decomposition by heterotrophic microorganisms; 3, animal excretion; 4, 5, colorless, green, and purple sulfur bacteria; 6, anaerobic sulfur-reducing bacteria, *Desulfovibrio*; 7, aerobic sulfide-oxidizing bacteria, *Thiobacilli*. Step 8 represents the conversion of phosphorus from an unavailable to an available form when iron sulfides are formed, illustrating how cycling of one vital element can affect another.

5. When iron sulfides are formed in the sediments, phosphorus is converted from insoluble to soluble form, as shown by the "phosphorus release" arrow (step 8 in Figure 6), and thus enters the pool available to living organisms. Recovery of phosphorus as a part of the sulfur cycle is most pronounced in the anaerobic (without oxygen) sediments of wetlands, which are also important sites for the recycling of nitrogen and carbon.

Our Leaky Suburbs

The Lake Wingra watershed (or catchment basin) in Wisconsin was the site of one of several landscape-level studies funded by the National Science Foundation as part of the United States' contribution to the International Biological Program in the 1970s. On one side of the lake is a large suburban development, while on the other side is a large forest preserve. At each spring thaw, accumulated fertilizers, soil, and sediments (including dog feces) leak out of the lawns into the lake, causing an intense algal bloom that turns the developed side of the lake a bright green. In contrast, very little of anything runs off from the forest, so that side of the lake remains clear (Watson et al. 1981). While the suburban landscape (like many other human-dominated and domesticated landscapes) leaks materials profusely, nutrients entering the forest preserve are recycled constantly with little net loss.

As more and more human-dominated landscapes leak both harmful and usable materials into our life-support systems, we need to devote more attention and technology to source reduction and recycling instead of after-the-fact waste treatment. More about this major shift in combating disorder created by pollution will be found in the Epilogue.

Resources Out of Place

Both the sulfur and nitrogen cycles are involved in urban and industrial air pollution. The **oxides** of nitrogen (N_2O and NO_2) and sulfur (SO_2) are toxic gases that normally are only transient steps in their respective cycles. The combustion of fossil fuels, however, has greatly increased the concentration of these volatile oxides in the air, not only in urban–industrial areas but also far downwind from such sites, to the point where they threaten the health of humans and of the green plants that support us. The burning of coal (especially in large power plants) is a major source of sulfur dioxide, and automobile exhaust, along with other industrial combustions, is a major source of nitrogen oxide. Together they constitute about one-third of the industrial pollutants discharged into the air over the United States and other industrialized nations.

Sulfur dioxide is damaging to photosynthesis, as was discovered by Haagen-Smit and associates in the early 1950s when leafy vegetables, fruit trees, and forests in the Los Angeles basin began to show signs of stress (Haagen-Smit et al. 1952). Furthermore, sulfur dioxide interacts with water vapor to produce dilute sulfuric acid (H_2SO_4) droplets that fall to earth as **acid rain**, a truly alarming development that has attracted the attention of the public and researchers throughout the world. Where soils and water lack pH buffers that neutralize acid, the acidity increases (the pH decreases) to levels that greatly stress vegetation and fish. The disappearance of fish from lakes in the Adirondacks in New York State and dying trees in the Black Forest of Germany are being blamed on acid rain, although it is not certain that acid rain is the only factor involved (ozone may also be involved, as will be noted in the next section). The building of tall smokestacks (Figure 7) for coal-burning power plants and other

FIGURE 7
Tall smokestacks reduce air pollution in the immediate area of the stack, but aggravate the problem of acid rain. (Photograph by J. N. A. Lott, McMaster Univ./Biological Photo Service.)

industries to reduce local air pollution has aggravated the problem, since the longer the oxides remain in cloud layers, the more acid is formed (which may also explain why mountain forests seem to suffer before those in the lowlands). This is a good example of a quick-fix or short-term solution that produces a more severe long-term problem.

The oxides of nitrogen, in the concentrations now occurring in the air, also threaten the quality of life. They irritate the respiratory membranes of higher animals and humans. They also contribute to acid rain, since nitrogen oxides combine with water to produce nitric acid. Furthermore, chemical reactions with other pollutants produce a **synergism** (where the total effect of an interaction exceeds the sum of the effects of each factor acting

Clean Air: Time to Act

There has been much discussion about controlling smokestack emissions from power plants, but widespread action has not yet been taken. The cost of emission control is so great that the economic damage done by pollutants has to be high before benefit–cost ratios for control and public "willingness to pay" are perceived to be positive. But waiting too long to take action could result in long-term damage that would cost billions in loss of health and life support. We must urge our political leaders to face up to the inevitable measures needed to reduce air pollution at its source.

There is some good news in a report that a "clean burning" coal power plant (where sulfur and other air pollutants are removed before coal is burned) operating in California is economically competitive with traditional plants (see Spencer et al. 1986). In other words, the technology is available, but so far (as of 1996) the Clean Air Acts passed by the U.S. Congress do not go far enough to require that this "know-how" be applied. In fact, going only part way can make things worse, since power plants can meet the current regulations simply by shifting to western low-sulfur coal, thereby putting eastern miners of high-sulfur coal out of business.

Again we come back to the concept of input management, where it is more effective economically in the long run to reduce pollution at its source, from the input side of the production system (the coal being burned), rather than from the output side (the effluent from the smokestack).

alone). For example, in the presence of ultraviolet light, nitrogen dioxide reacts with unburned hydrocarbons (both are produced in large quantities by automobiles) to produce **photochemical smog** (chemically, peroxyacetyl nitrate and ozone, both classified as "photochemical oxidants"), which not only makes one's eyes tear but is generally dangerous to one's health. Hall et al. (1992) have estimated that people living in the California south coastal basin average 17 days' absence from work per year due to air pollution-related sickness, and that the economic value of avoiding these effects would be nearly 10 billion dollars. One need only go into the hills of southern California, the Appalachians, or West Germany to see first-hand the damage to pine forests caused by smog. Yellowing and early loss of needles, dieback of tree crowns, reduction in growth, and ultimately, tree death are symptoms of the extreme stress produced by photochemical oxidants. Damage to crops is not so obvious, but it is nevertheless extensive. It is estimated that a 15 percent reduction in bean production and a total annual loss of 45 million dollars in vegetables results from oxidant air pollution in southern California (Kneese 1984).

Ozone, a "Chemical Weed"

A weed is sometimes defined as a plant in the wrong place—that is, a generally useful or harmless plant that insists on growing in a place where you don't want it (such as your garden). Resources that are essential to life when in their normal or natural position in cycles can cause trouble when their amounts are increased or when they turn up in the wrong place as a result of human activities. Ozone (O_3) is a prime example of something that we cannot live without, yet when in the wrong place is a costly and dangerous "chemical weed." Ozone is formed naturally in the stratosphere as incoming solar radiation interacts with oxygen. As discussed in Chapter 4, the ozone layer in the upper atmosphere shields us from deadly ultraviolet radiation. Certain chemical pollutants can break down this life-sustaining shield, a prospect so frightening that the industrial nations have taken steps to decrease their production.

But at the same time that we strive to maintain ozone in its proper place, ozone in the *lower* atmosphere is becoming a major photochemical oxidant pollutant at ground level. One study showed that ozone, in the concentrations of 0.02–0.14 ppm (parts per million) that now exist even in areas far removed from large cities, reduced photosynthesis in all species of crops and trees tested (Reich and Amundson 1985). Such findings suggest that ground-level ozone may be a greater threat to us and our life-support systems than acid rain. Or, as we might expect, there could be a synergism between the two. Kneese (1984), in a study of the economic benefits of clean air and water, calculated that even a very small reduction of 0.01 ppm in ground-level ozone concentration would result in a million fewer cases of chronic respiratory disease in the work force, yielding a benefit to business of greater than a billion dollars a year.

The Global Carbon Cycle

Figure 8 is a model of the global carbon cycle. It includes estimates of the amounts of carbon dioxide (CO_2) in four major compartments: atmosphere, oceans (including uplifted carbonate sediments such as the famous white cliffs of Dover), terrestrial biomass, and soils and fossil fuels. Flux rates between compartments are shown by the arrows. The atmospheric pool is small in comparison with the amounts in the other compartments, but it is a very active pool that is being increased by the burning of fossil fuels and the clearing and plowing of land for agriculture—both of which release carbon dioxide into the air. Prior to the industrial age, it is believed, flows between atmosphere, continents, and oceans were bal-

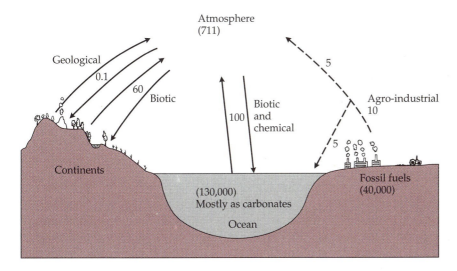

FIGURE 8

The global carbon cycle. Figures are based on estimates of 10^9 tons in major biosphere compartments and circulating between compartments (arrows). The small atmospheric compartment is being increasingly affected by carbon dioxide released by human activities.

anced, as shown by the solid lines in Figure 8. During the past century, however, the carbon dioxide content of the atmosphere has been slowly rising, because anthropogenic inputs exceed the removal capacity of the vegetation and the ocean's carbonate system (as shown by the dashed lines in Figure 8). The atmospheric concentration has risen from about 290 ppm (0.29 percent) in the early 1800s (estimated, since there were no measurements then) to 315 ppm in 1958, when accurate measurements were first made, to 345 ppm in 1990—and it is still rising.

The Uneasy Balance between Global Warming and Cooling

While more carbon dioxide in the air could have the positive effect of increasing primary production, it raises concern about possible undesirable changes in climate due to a **greenhouse effect**. Carbon dioxide in the atmosphere acts like glass in a greenhouse; it lets in the short waves of sunlight but reflects the long waves of heat reradiated by the earth, thus retarding the outflow of heat from the biosphere. Should the present rate of increase in carbon dioxide and other greenhouse gases, such as methane, continue into the next century, a global warming could occur. It would take a rise in mean temperature of only 1–4 degrees Celsius to bring about a rapid rise in sea levels due to the heating of the ocean (warm water occupies a larger volume than cold water) and polar ice melts; if that happened, we could say goodbye to New York and other coastal cities. A warming could also cause changes in rainfall patterns that could disrupt agricultural production, which might be an even worse disaster.

Before you take to the hills, we hasten to point out that contrary forces are operating that tend to cool the earth. In recent geological times, ice ages have occurred about every 10 to 20 thousand years (and it has been about that long since the last one, so we could be ripe for another). Natural events such as volcanic eruptions, as well as human activities, put particulate matter (dust and other airborne particles) into the atmosphere, which reflects incoming solar radiation and thus cools the earth. In recent years a "pollutant haze" of microscopic particles from smokestacks and other sources has thickened over industrial regions of eastern North America, Central Europe, and Eastern Asia to such an extent that air temperatures have actually decreased in those areas, in sharp contrast to most of the rest of the earth (see Kerr 1995). The increase in cloud cover that may accompany an anthropogenic rise in temperature would also counteract global warming. However, these possible benefits by no means balance the costs of air pollution. Both haze and cloud cover reduce sunlight reaching the earth's surface, which reduces primary production, and the haze is bound to contain toxic materials. A nuclear war could throw up so much dust that what is widely discussed as "nuclear winter" would occur (Ehrlich et al. 1983). A cluster of meteorites striking the earth could do the same thing. With so many possibilities for both cooling and heating, it is imperative that we increase international efforts to monitor changes in temperature, atmospheric carbon dioxide, comet trajectories, and other threats to global balance so that we will have a better idea of what needs to be done to avoid undesirable climatic changes. (For more on the greenhouse effect and possible global climatic changes, see Bolin et al. 1986; Gates 1993.)

Other Atmospheric Gases

In addition to carbon dioxide, two other forms of carbon are present in the atmosphere in smaller amounts: carbon monoxide (CO), at about 0.1 ppm, and methane (CH_4), at about 1.6 ppm. These flux rapidly and thus have short residence times—about 0.1 year for carbon monoxide and 3.6 years for methane, as compared with 4 years for carbon dioxide. Both carbon monoxide and methane arise from incomplete or anaerobic decomposition of organic matter; in the atmosphere both are oxidized to carbon dioxide. An amount of carbon monoxide equal to that released by natural decomposition is now injected into the air by the incomplete burning of fossil fuels, especially automobile exhaust. Carbon monoxide, a deadly poison to humans, is not a global threat, but it is a worrisome pollutant in urban areas when the air is stagnant. Concentrations as high as 100 ppm are not uncommon in areas of heavy automobile traffic. (Pack-a-day cigarette smokers receive up to 400 ppm, which appreciably reduces the oxygen-carrying capacity of their blood.)

Methane, which is produced in quantity by wetlands and also by cows and termites (whose digestive systems are anaerobic), is another atmospheric component with the potential for both positive and negative effects. It is believed to have a role in maintaining the stability of the ozone layer in the upper atmosphere. But too much methane can be disruptive to heat balances.

Nutrient Cycling in Nutrient-Poor Soils

One of the myths about the tropics is that the soils there are fertile and capable of feeding the world if we would just remove the forests and plant crops. There are, of course, areas of fertile soil in the warmer climates, but soils over huge areas, such as the tropical rain forests of the Amazon basin, are quite poor compared with areas such as the prairie soils of Iowa (Jordan 1985). Luxuriant forests are able to persist in the Amazon only because of efficient biotic recycling mechanisms that keep vital nutrients such as phosphorus and nitrogen circulating within the biomass. In such forests, less than half of the available pool of nutrients is in the soil, as compared with more than 90 percent in European or eastern North American forests. When vegetation is removed from temperate forests or prairies for agricultural purposes, the soils retain their nutrients and structure. They can be conventionally farmed for many years, which involves plowing one or more times a year, planting short-season annual plants, and applying large amounts of quick-release inorganic fertilizers. During the winter, freezing temperatures help hold in nutrients and combat pests and diseases. In the tropics, on the other hand, forest removal takes away the land's ability to hold and recycle nutrients (as well as to combat pests) in the face of high year-round temperatures and periods of leaching rainfall. The thin tropical soils lack organic and biotic holding mechanisms, so any nutrients left in them are quickly drained away (see Figure 15 in Chapter 8). Crop production declines rapidly (sometimes after only 2 or 3 years), and the land is abandoned, creating the pattern of **shifting agriculture** (also called **swidden agriculture**) so common in the tropics.

Among the biotic devices that aid in keeping nutrients cycling within the living biomass in tropical forests are the following:

1. **Root mats** consisting of many fine feeders penetrating the surface litter quickly recover nutrients from fallen leaves before they can be leached away. Root mats apparently also inhibit the activities of denitrifying bacteria, thus blocking the loss of nitrogen to the air. Some tropical trees even have "upwardly mobile roots" that grow upward on the tree trunks (instead of downward into the soil as do normal roots) and are thus able to absorb nutrients from rainwater flowing down the stem (Sanford 1987).
2. **Mycorrhizal fungi**, symbiotic microorganisms associated with root systems, act as nutrient traps, greatly facilitating the recovery of nutrients and their retention within the biomass. (This symbiosis between higher plant and microorganism for mutual benefit is widespread on poor soils in the temperate zone as well, as will be shown in Chapter 6.)
3. **Evergreen leaves** with thick, waxy cuticles and thick bark retard loss of water and nutrients and also resist herbivores and parasites.
4. **Algae** and **lichens** that cover the surfaces of many leaves scavenge nutrients from rainfall and fix nitrogen from the air.

(For more on nutrient cycling in tropical forests, see Jordan 1982, 1985.)

This brief account, of course, oversimplifies complex situations, but it shows why some sites in the tropics that support luxurious forests yield so poorly under northern-style crop management. It is evident that a different type of agriculture needs to be designed for the tropics—one involving reduced soil disturbance (less plowing), more perennial plants that use C_4 photosynthesis and perhaps mycorrhizae, more multiple cropping, and more use of legumes and other nitrogen fixers.

Unfortunately, agronomic science is so locked into industrialized agriculture concepts, and agricultural economics is so dependent on annual grain cash crops, that the necessary redesign of agriculture is slow in developing. We need, especially, to take a good look at systems of traditional agriculture developed and sustained for centuries by tropical peoples themselves. These include paddy and flood-irrigated rice culture, mixed cultures such as the corn–beans–squash crops developed by the ancient Mayans and still in use today in Mexico, horticultural systems involving food-bearing trees and shrubs mixed in with annual vegetable and grain crops, and combination crop and Chinese fish culture (in which crop residues are fed to fish) (Gliessman et al. 1981; Altieri 1987). Energy-based comparisons of traditional and industrial agriculture will be presented in Chapter 8.

Recycling Pathways

Since we are concerned more and more with recycling problems, both in nature and in commerce, it is instructive to review the subject of biogeochemistry in terms of recycling pathways. Figure 9 shows a number of routes by which resources are recycled. As already indicated, the recycling of many vital nutrients involves microorganisms and energy derived from the decomposition of organic matter (pathway 1 in Figure 9). Where small plants such as grass or phytoplankton are heavily grazed, recycling by way of animal excretion may be important (path 2). In nutrient-poor situations, a direct return (path 3) is accomplished by symbiotic microorganisms that become a part of autotrophs (plants), such as the mycorrhizal fungi mentioned in the preceding section. As was shown in the diagram of the water cycle (Figure 1), many substances are recycled by physical means involving solar energy (path 4). Finally, fuel energy is used by humans to recycle water, fertilizers, metals, and paper (path 5). Note again that recycling requires energy dissipation from some source, such as organic matter (paths 1, 2, 3), solar energy (path 4), or fuel (path 5).

Recycling of Paper

Paper provides a good example of how recycling develops in urban-industrial systems in a manner parallel to the recycling of important materials in natural ecosystems. In either case, energy expenditure for recycling becomes desirable or necessary when resources become scarce or waste products pile up to the detriment of life within the system.

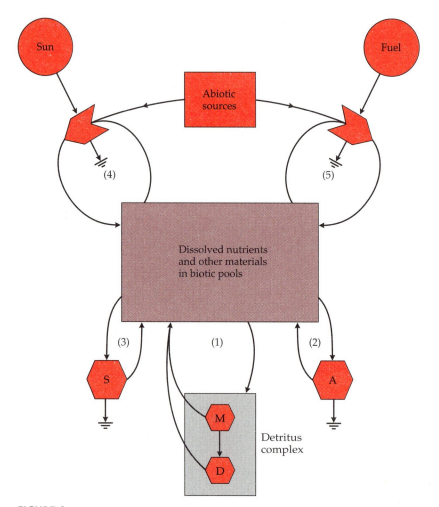

FIGURE 9
Five major recycling pathways: (1) decomposition (M, microorganisms; D, detritus consumers); (2) animal (A) excretion; (3) symbiotic organisms (S); (4) solar energy-driven, as in the hydrological cycle; (5) fossil fuel-driven, as in industrial recycling.

As long as there are plenty of trees and paper mills, and plenty of vacant land for landfills, there is little incentive to invest in the facilities and energy (for pickup, sorting, and transport of waste paper) needed for the establishment of an efficient recycling program. This situation is shown in Figure 10A. But as the environs of the city become congested, land values rise, and it becomes increasingly difficult and expensive to maintain landfills or to find new sites as old ones fill up. Or the cost of new paper may increase because pulpwood supplies or mill production fall short of demand (as is now the case in much of Europe). In both cases it pays to consider recycling. For recycling to be successful there must be a recycling mill that provides a market for used newspaper and cardboard, as shown in Figure 10B. Such a mill is comparable to the mycorrhizal recycling system of the tropical rain forest. In other words, some-

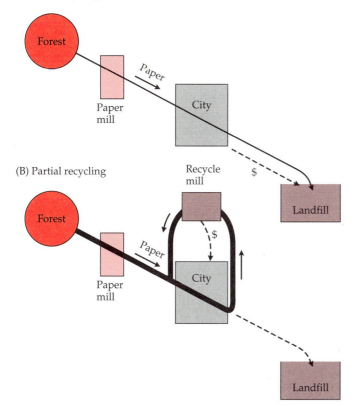

(A) Zero recycling

(B) Partial recycling

FIGURE 10
Conditions not conducive (A) and conducive (B) to recycling of paper. Benefits of recycling for the public as a whole are that harmful environmental impacts (on forests, streams, and land) and taxes for city services are reduced. Requirements for recycling include: citizen participation, a collection system (modified garbage pickup), a sorting and packaging warehouse, a recycling mill, transportation to the mill, a market for used paper, and profit for the city or cost less than that of a landfill.

thing new has to be added to the total system. Also, as shown by the dotted lines representing money flow in Figure 10B, the city, which previously had to pour money into maintaining the landfill, now gets an income from selling used paper, which helps pay for the cost of disposal of other wastes.

Limiting Factor Concepts

The idea that organisms may be controlled by the weakest link in an ecological chain of requirements goes back a century or more to the time of Justus Liebig, who was a pioneer in the study of inorganic chemical fertilizers in agriculture. Liebig observed that the growth of crop plants was often limited by whatever essential element was in shortest supply (in terms of requirements), regardless of whether the amount required was large or small. **Liebig's law of the minimum** has come to mean that growth is limited by that nutrient that is least available in terms of need. The idea can be extended to include factors other than nutrients and to include the limiting effect of the maximum (that is, too much can also limit). We must also recognize that factors interact; that is, a short supply of one thing may affect requirements for another thing that is not in itself limiting. The result is a useful working principle.

We may restate the extended **concept of limiting factors** as follows: The success of an organism, population, or community depends on a complex of conditions; any condition that approaches or exceeds the limit of tolerance for the organism or group in question may be said to be a limiting factor. The chief value of the limiting factor concept is that it gives the ecologist an "entering wedge" into the study of complicated situations. Environmental relations are indeed complex, so it is fortunate that not all factors are of equal importance in a given situation. Oxygen, for example, is a physiological necessity for most animals, but it becomes a limiting factor only in environments where it is in short supply relative to demand. If fish are dying in a stream receiving sewage, for example, oxygen concentration in the water would be one of the first things we would investigate; oxygen in water is variable, easily depleted by decomposition, and often in short supply. If small mammals are dying in a field, however, we would look for some other cause; oxygen in the air is constant and abundant in terms of need (not easily depleted by biological activity) and, therefore, not likely to be limiting to air-breathing animals living above ground.

Liebig's law (and the limiting factor concept in general) is most applicable to stable conditions where inflows balance outflows, and least applicable under transient-state conditions, where flows are unbalanced and where rates of function will likely depend on rapidly changing concentrations and the interactions of many factors. In a recent assessment of resource limitation, Bloom et al. (1985) concluded that plants are able to adjust their allocation of resources so that their limits of growth become nearly equal for all resources. They argued that this theorem describes growth in relation to resource limitation more accurately than does Liebig's law. In any case, there are a handful of natural resources that can be singled out as frequently being major limiting factors for ecosystems and humans alike: water on land, oxygen in water, and nitrogen, phosphorus, and energy in many environments, as already discussed.

Since there is no theoretical basis for any "one-factor" control hypothesis under transient-state conditions, the strategy of pollution control must involve reducing the input of as many enriching and toxic materials as possible, not just of one or two items.

Factor Compensation

Species with wide geographic ranges often develop locally adapted genetic races or subpopulations, called **ecotypes**, with different growth forms or different limits of tolerance for temperature, light, nutrients, or other factors. Use of reciprocal transplants can reveal whether compensation along a gradient of conditions has led to different genetic races or is due merely to acclimatization by individuals. The possibility of genetic fixation in local strains has often been overlooked in applied ecology, with the result that restocking has often failed because individuals from remote regions, not adapted to the local conditions, were used to restock a depleted ecotype.

A good example of temperature compensation without genetic fixation within a species is shown in Figure 11A. The small jellyfish *Aurelia aurita* moves through the water by rhythmic contractions that expel water from its central cavity in a sort of jet propulsion. A pulsation rate of 15–20 per minute seems to be the optimum. Individuals living in the northern sea near Halifax, Nova Scotia, swim at the same optimum rate as individuals in southern seas, even though water temperatures are 15–20 degrees lower.

Figure 11B illustrates true ecotypes of a species of yarrow that ranges all the way from sea level to high altitudes in the Rocky Mountains. As shown in the diagram, high-altitude plants retain their short stature, and plants of the low-altitude race their tall stature, when both are grown from seed in the same garden, indicating that genetic fixation has taken place.

Many species have narrow ranges of tolerance and are, accordingly, sensitive to change. Such species can be useful **ecological indicators** of changes in environmental conditions. Range managers, for example, find that the relative abundance of plants that are sensitive to grazing can indi-

(A) Jellyfish (*Aurelia aurita*)

(B) Yarrow (*Achillea millifolium*)

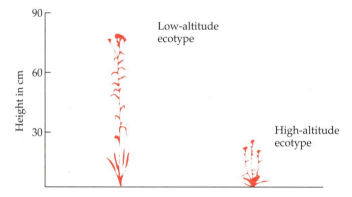

FIGURE 11
Physical factor compensation in animals and plants. (A) Different geographic populations of the same species of jellyfish are acclimated to swim at about the same rate despite the different water temperatures of their environment. (After Bullock 1955.) (B) When high-altitude and low-altitude individuals of a species of yarrow are grown from seed in the same garden at sea level, they retain their tall or short stature, indicating that genetic fixation has taken place. The two varieties are thus true ecotypes. (After Clausen et al. 1948.)

cate the approach of overgrazing before it becomes apparent in the grassland as a whole. Species that decline in abundance with grazing are called "decreasers," and species that increase with grazing, perhaps because they are spiny or taste bad, are called "increasers." Likewise, indicator species (both plants and animals) often prove useful in the assessment of different kinds of water pollution.

Biological Clocks

Organisms not only adapt to the physical environment, but also use natural periodicities in the physical environment to time their activities and to "program" their life histories so they can benefit from favorable conditions. They accomplish this by means of **biological clocks**, physiological mechanisms for measuring time. The most common and perhaps basic manifestation is the **circadian rhythm** (*circa*: "about"; *dies*: "day"), or the ability to time and repeat functions at about 24-hour intervals even in the absence of conspicuous environmental clues such as daylight. It is this rhythm that gets upset when we suffer "jet lag" after long airplane trips. The biological clock is set by environmental and physical rhythms and enables organisms to anticipate daily, seasonal, tidal, and other periodicities. Physiologically, the timing device involves a "circadian axis" of interaction between the eye, a lower brain center, and, in vertebrates, the pineal gland, which secretes the hormone melatonin, now being marketed as a sort of natural sleeping pill. There is also apparently a second timing device that actually measures time like an hourglass or stopwatch. (For a review of theories see Morell 1996.)

A dependable clue used by organisms to time their seasonal activities in the temperate zone is length of day or **photoperiod**. In contrast to temperature and most other seasonal factors, photoperiod is always the same in a given season and locality. The seasonal variation in the photoperiod increases with increasing latitude, thus providing latitudinal as well as seasonal clues. At Winnipeg, Canada, for example, the maximum photoperiod is 16.5 hours (in June) and the minimum is 8.0 hours (in late December). In Miami, Florida, the range is only 13.5 (June) to 10.5 hours (December).

Photoperiod has been shown to be the timer or trigger that sets off physiological sequences that bring about the growth and flowering of many plants; molting, fat deposition, and migration in birds and mammals; and the onset of diapause (resting stage) in insects. Day length is sensed through a receptor such as the eye in animals or a special pigment in the leaves of a plant. This, in turn, activates one or more hormone or enzyme systems that bring about the associated physiological or behavioral response. Although higher animals and plants are widely divergent in morphology, their response to day length is remarkably similar. In many, but by no means all, photoperiod-sensitive organisms, timing can be altered by experimental or artificial manipulation of the photoperiod. Florists, for example, can often force flowers to bloom out of season by altering day length in the greenhouse.

In striking contrast to day length, rainfall in a desert is highly unpredictable. Yet desert plants adapt to this uncertainty in a unique way. The seeds of many desert annuals contain a germination inhibitor that must be washed out by a certain minimum amount of rain (for example, one-half inch or more). This much rain provides all the water necessary for the plant to grow, produce seeds, and complete its life cycle. If such seeds are placed in moist soil in a greenhouse, they fail to germinate, but they do so quickly when treated with a simulated shower of the necessary magnitude. Seeds of desert flowers may remain viable in the soil for many years, waiting for an adequate shower, which explains why deserts "bloom" so quickly after a heavy rainfall.

Fire as an Ecological Factor

Fire is a major environmental factor that is almost a part of climate in shaping the history of vegetation in most terrestrial environments of the world. It is especially important in forest and grassland regions of the temperate zone and in tropical regions with dry seasons. In most parts of the United States, especially in the southern and western states, it is difficult to find a sizable area of forest or grassland that does not give evidence of fire having occurred there in the past 50–100 years. Fire was a factor in natural ecosystems long before modern times. Fires are started naturally by lightning, and early humans (such as the North American Indians) regularly burned woods and prairies to flush game or open up areas for agriculture and grazing. Contrary to popular opinion, fire in modern times is not necessarily detrimental. **Controlled burning** (also called prescribed burning), as we shall see, can be a very useful tool in the management of certain types of forests and grasslands.

In regions where seasonal or periodic minor fires are frequent, we find **fire-adapted vegetation** whose prosperity or very survival depends on fire. The chaparral (dwarf evergreen shrub forest) of southern California that was discussed in Chapter 2, the long-leaf pine forests of the southeastern United States, and the plains of East Africa where the great herds of antelope roam, are three well-studied examples of fire-adapted vegetation. How fire acts to maintain a southwestern grassland is shown in Figure 12. In this case the grass not only is adapted to fire, but is more valu-

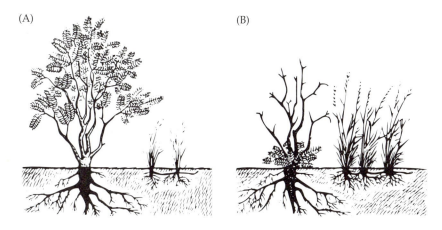

FIGURE 12
Diagrams show how fire favors grass over mesquite shrubs in the southwestern United States. (A) In the absence of fire, the mesquite chokes out the grass. (B) After a fire, grass recovers quickly, growing with increased vigor under conditions of reduced competition. Controlled burning will eliminate the mesquite entirely and maintain the grassland. (After Cooper 1961.)

able for grazing than the shrubs that tend to increase in the absence of fire. Controlled burning can help maintain the grassland against encroachment by shrubs.

In dry or hot regions, fire acts as a decomposer, releasing mineral nutrients from accumulated old litter that becomes so dry that the bacteria and fungi of decay cannot act on it. In such cases fire may actually increase productivity by speeding up recycling. Furthermore, periodic light fires prevent the start of severe fires by keeping the combustible litter to a minimum. After several decades of adhering to a strict fire suppression policy in our national parks (Yellowstone, for example) and forests, foresters are now turning to controlled burning as a means of preventing disastrous wildfires (Oberle 1969). Whether destruction of California homes by "firestorms" could be prevented by a program of controlled burning is a significant question.

The Soil Resource

Air (the atmosphere), water (the hydrosphere), and soil (the pedosphere) are the media that support life. Air and water have been featured so far in this chapter; soil is the third major life-support component of the bios-

Smokey Bear is (Partly) Right

It is extremely important that we distinguish between the light, largely surface fires characteristic of fire-adapted ecosystems and the wild, destructive crown forest fires that make headline news. Because humans, through carelessness, tend to create such holocausts, it is necessary that the public be made sharply aware of the necessity of fire prevention—hence the Smokey Bear posters. The citizen must realize that he or she should never cause fires anywhere in nature. However, it should be understood that fire, often started by lightning, is part of the natural environment in many regions and that the scientific use of fire as a tool by trained persons is in the public interest.

phere, and human activities are affecting this vital resource as well. **Soil** is the product of the physical weathering of the earth's crust (e.g., rocks and clay minerals) and the activities of organisms, especially vegetation and microorganisms. Soil is being viewed more and more not just as a mixture of clay, sand, and humus in which plants grow, but as a living system. More attention is being paid to the incredible diversity of life in the soil, including not only the microbiota (bacteria and fungi) but also soil animals (mites, collembolans, earthworms) and the metabolically active root zones (rhizospheres) (see, for example, Coleman and Crossley 1996).

For the terrestrial ecosystem and landscape, soil functions as the chief organizing center, since all ingredients necessary to sustain primary and secondary production are stored in and recycled by the soil (the "brown belt" mentioned in Chapter 1). Accordingly, soil may be the best indicator of overall environmental health. The theory is that if the quality of the soil is being maintained, then whatever is going on above ground (natural vegetation, agriculture, forestry) is sustainable. The problem is that there is as yet no consensus as to how to measure and monitor soil quality. Obviously, some kind of multiple factor index is needed. There is a large literature on this subject containing many suggestions (see especially Paul 1989 and the Soil Science Society of America 1994).

The cut edge of a bank or trench, as shown in Figure 13, reveals that soil is composed of distinct layers, which often differ in color. These layers are called **soil horizons**, and the sequence of horizons from the surface down is called a **soil profile**. The upper or **A horizon (topsoil)** is composed of bodies of plants and animals reduced to finely divided organic matter, mixed with clay, sand, silt, and other mineral matter. The second or **B horizon** is composed of mineral soil in which organic matter is mineralized (decomposed to inorganic compounds) and thoroughly mixed with finely divided "parent" material. The soluble materials in the B horizon are often formed in the A horizon and moved downward (leached) by water flow. The dark band in Figure 13 represents the upper part of the B horizon where such materials have accumulated. The third or **C horizon** is the more or less unmodified parent material. The **parent material** is the original geological material that is disintegrating in place, or material that has been transported to the site by gravity, glaciers, wind, water (alluvial deposit), or volcanic action. Soils developing on transported materials are

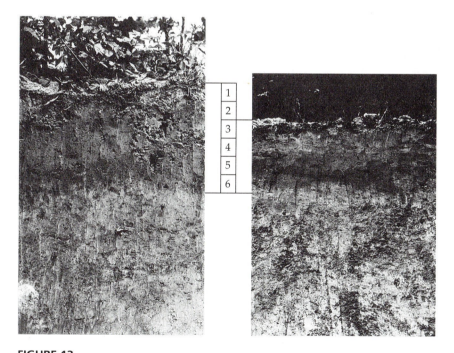

FIGURE 13

Soil profile in a virgin area compared with an eroded area within an eastern deciduous forest region. The dark layer (1–6) is the A horizon or topsoil. The lighter area below is the B horizon where leached material accumulates. Note that the eroded area has lost one-third of its topsoil. (Photographs courtesy of Soil Conservation Service.)

often very fertile, as seen in the deep wind-deposited soils of Iowa, the rich soils of the deltas of large rivers, and the fertile soils that develop on volcanic ash.

In terms of nutrient cycling, the organic portion of soil can be divided into two pools. The first is a labile (active) fraction, the immediate source of nutrients for plants. Microorganisms constitute a considerable part of this fraction. The second is a stable, less active fraction composed of humic substances, the "humus capital" referred to in Chapter 4. This pool is resistant to decay and provides the texture that is so important in maintaining soil quality (Paul 1989).

Soil Development

With the passage of time, soils go from youth to maturity to old age, a transition analogous to the development of an organism or a biotic community. Youthful soils (like young organisms) rapidly accumulate organic matter and develop structure (profile), as shown in Figure 13. Maturity is reached with the achievement of an approximate steady state (gains equal losses over the long run). On favorable substrata this stage may be reached in as little as 100 to 2000 years from the time the parent material was first exposed (time zero), depending on climate, vegetation, and other conditions (Hall et al. 1982). Development of thin soils (shallow soils over rock, sand, or other limiting layers) is, of course, very slow. As soils age (after thousands of years), nutrients may leach out faster than they accumulate, and impervious layers (hardpans) may form that restrict root, air, and water penetration. From an agricultural standpoint, soils tend to be most fertile early in their development, with maximum potential reached after a certain level of organic matter accumulation and structural development has taken place, but before extensive weathering and the development of restrictive horizons.

Major Soil Types

The ten major soil types (orders) currently classified by soil scientists are listed in Table 1. The types are arranged in order of the percentage of the terrestrial world they occupy, and for comparison, the percentage of the United States they occupy is also shown. Entisols are the most youthful and ultisols the most weathered. Alfisols (moderately weathered forest soils) and mollisols (grassland soils) make the best agricultural soils. These constitute only about 24 percent of land area worldwide, but about 38 percent of the continental United States, which has less area in desert and weathered tropical soils (aridisols and oxisols) than the world aver-

TABLE 1   Distribution of Major Soil Types Worldwide and in the United States

Soil Type	Percent of Land Area	
	Worldwide	United States
Aridisols (desert soils)	**19.2**	**11.5**
Inceptisols (weakly developed soils)	15.8	18.2
Alfisols (moderately weathered forest soils)[a]	14.7	13.4
Entisols (recent soils, profile undeveloped)	12.5	7.9
Oxisols (tropical soils)	9.2	0.02
Mollisols (grassland soils)[a]	9.0	24.6
Ultisols (highly weathered forest soils)	8.5	12.9
Spodosols (northern conifer forest soils)	5.4	5.1
Vertisols (expandable clay soils)	2.1	1.0
Histosols (organic soils)	0.8	0.5
Miscellaneous soils (steep mountains, etc.)	2.3	4.88
Total	100.0	100.00

Data from Steila 1976 and USDA 1975.

[a]Alfisols and mollisols are the best agricultural soils; together they cover 24% of world land area, 38% of United States land area.

age. The fact that three-fourths of the world's soils are unsuitable for intensive crop production unless heavily amended with fertilizer and water was commented on in the discussion of the world food crisis in Chapter 4.

Soil Displacement: Natural and Human-Accelerated

Soil erosion caused by water and wind occurs naturally at low rates all the time, with periodic large displacements resulting from great floods, glaciers, volcanic eruptions, comet impacts (perhaps), and other episodic events. Areas that lose soil faster than new soil is formed generally suffer reduced productivity and other detrimental effects. Areas receiving too much soil may also be negatively impacted, as in the case of the Illinois River noted in Chapter 1. However, fertility may be enhanced when soils are washed down from hills into river valleys and deltas, or are deposited on prairies by wind, as already noted. As is the case for so many natural processes, humans tend to accelerate soil erosion, often to our long-term detriment.

In the 1930s, the Soil Conservation Service (SCS) was established by the U.S. government to combat the soil erosion that was ruining thousands of acres of farm and forest land. The "dust bowl" was taking its toll on the western plains at about this time. The program that was developed to save soil is an excellent example of how government should work in the public interest in a democracy. A close linkage was established between the federal government in Washington, D.C., state governments, land-grant universities, and counties. Washington provided the money, and the universities contributed the research, but decisions were made locally, and county agents worked directly with landowners. Terracing, grass waterways, riparian forest buffer strips (Lowrance et al. 1984), crop rotation, and other measures, together with improvements in the economic and educational status of farmers, reversed the tide of soil loss, and a **soil conservation ethic** became generally accepted by farmers and other landowners.

Perhaps partly because of its success, the SCS had so much support in Congress and in the states that it became increasingly bureaucratic (i.e., less responsive to real needs) and extended its activities into other areas, such as channeling streams and building large dams (Figure 14), that often had questionable value in soil preservation. Then, suddenly, in the 1970s, soil erosion again became an urgent national problem because of two new trends. The first is the industrialization of farming, emphasizing cash crops that are treated less as food than as commodities for sale, especially on the overseas market. Unfortunately, when farms are operated strictly as businesses, often by corporations or other absentee owners, crop yield in the short term is maximized at the expense of maintenance of long-term fertility and productivity. The second trend is urban sprawl, as roads and housing developments mushroom into the rural countryside with little or no concern about the loss of soil and prime farmland.

The urgent need to counteract the deleterious effects of these two major land-use changes and reestablish a soil conservation ethic is well documented by governmental reports (for example, Council on Environmental Quality 1981) and assessments by the private Conservation Foundation (Batie and Healy 1983; Batie 1983; Clark et al. 1985). The Soil Conservation Service considers the maximum "tolerable" level of annual soil loss from good, deep soils to be 5 tons per acre, and from poorer, thinner soils to be 2 tons per acre. According to the surveys just cited, half of the best farmland in Iowa and Illinois is losing 10–20 tons per acre each year, and a quarter of all farmland in the United States is losing soil at a rate greater than the "tolerable" level. To put this in perspective, consider that an acre of good topsoil 6 inches deep (about plow depth) weighs about 1000 tons, so 1 acre-inch equals about 167 tons. An annual loss of 10 tons per acre results in a loss of 1 inch of topsoil every 17 years—a loss much greater

(A)

(B)

FIGURE 14

(A) Reservoir abandoned because of excessive silting just a few years after construction of the expensive dam. Although watershed fires and "unusual" rains were blamed for the erosion, an ecosystem-level study of the sparsely vegetated watershed would probably have shown that this was no place to build a dam. (B) Inexcusable soil erosion during construction of an office park. A farmer's future welfare depends on reducing soil loss; the developer has no such incentive unless he is faced with large fines for the kind of devastation shown in this picture. (Photographs courtesy of Soil Conservation Service.)

than any known rate of soil formation. Langdale et al. (1979) estimate that for every inch of topsoil lost, a crop yield reduction of at least 10 percent occurs. Soil losses from urban and suburban construction, although often of short duration, are even more severe (Figure 14B). Losses of 40 tons per acre are not uncommon, and 100 tons per acre have been recorded in extreme cases (Clark et al. 1985).

Soil erosion resulting from poor land use is, of course, not new. Even a cursory study of history reveals that misuse of land and associated loss of topsoil has been a leading cause of the decline of many past civilizations (Carter and Dale 1974). What are new are the accelerated rate and scale of soil disturbance due to market pressure, population increase, and use of large, powerful machines; and the toxic agricultural and industrial chemicals that move downhill and downstream with the displaced soil. If the current degradation continues, our needs and demands for more food from fewer acres cannot possibly be met.

Sustainable Tillage

Fortunately, the seriousness of the modern threats to our life-supporting soil is being recognized worldwide. New methods of farming that build up rather than use up soil are rapidly coming into use. One example is **conservation tillage** (also called **residue management**), which involves keeping a plant cover and/or a mat of plant residue on the soil at all times, and reducing plowing (**limited till**) or eliminating it altogether (**no-till**). An example of this method is shown in Figure 15. Not only does conservation tillage greatly reduce erosion or stop it altogether (Langdale et al. 1991), it improves soil quality and can sustain yields that are equal to or better than those of conventional tillage. Water and fertilizer retention are increased, and runoff of water and pesticides is decreased. Reducing the soil disturbance caused by frequent plowing and compaction by heavy machinery allows natural soil-building organisms and processes to operate. At first, it was thought that more herbicides would be required to control weeds in the absence of plowing, and this has occurred in some cases; however, recent field experiments have indicated that this need not be the case. For more on conservation tillage, see the reviews by Phillips et al. (1980) and Gebhardt et al. (1985), and the book by Little (1987).

Doing away with the moldboard plow (the kind that turns the topsoil upside down) is by no means a new idea. In 1943, Edward H. Faulkner published a small book, *Plowman's Folly*, based on his experiments with no-till farming. After serving many years as a county agent and teacher of agriculture, he became convinced that there was no scientific reason for plowing, and he set out to prove his convictions on his own farm in Ohio.

FIGURE 15
Example of conservation tillage in Iowa. Soybeans have been planted in the mulching residue of last year's corn crop without any plowing (a no-tillage system). (Photograph courtesy of Soil Conservation Service.)

Faulkner's methods were a radical departure from traditional farming at the time, but have proved to be remarkably similar to procedures now widely accepted as the wave of the future (Sidey 1992).

Ultimately, the fate of the soil system depends on society's willingness to intervene in the marketplace and to forgo some of the short-term benefits that accrue from "mining" the soil so that soil quality and fertility can be maintained over the longer term. The technology to do this is already available and will improve with more emphasis on agroecological research. Soils can be rebuilt but, as always, preservation is much cheaper.

An often overlooked and poorly regulated source of land deterioration and soil erosion is mineral mining operations. By one estimate more material is stripped from the earth by mining than by all natural erosion (Young 1992). The damage is especially severe in developing countries, which produce a large portion of the world's mineral supplies but use a relatively small share. As is too often the case, we in the developed countries prosper while the poor countries not only do not share equally in the economic returns but also suffer disproportionate environmental damage.

> ### Land Degradation Worldwide
>
> In 1992 land degradation, estimated as a percentage of vegetated land damaged, was as follows: worldwide, 17 percent; Europe, 23 percent; Asia, 20 percent; South America, 14 percent; and North America, 5 percent (World Resource Institute 1992–1993). It is not clear whether these figures include land devastated by mining. Deforestation is a major cause of soil loss in Asia and South America, but not yet in North America. However, if the current clear-cut mania, born of short-term economics, continues in North America, deforestation could quickly become a major factor. All these data suggest that intense land use over long periods of time, as in Europe and Asia, is the major contributing factor in land degradation. I would suggest that the New World has the opportunity and the obligation to show how progressive deterioration can be halted as land use intensifies in this part of the world.

There are many "Copperhills" (see Figure 12 in Chapter 3) in the less developed world.

Toxic Wastes: The Bane of Industrial Societies

Our life-support environment exhibits considerable ability to recover from periodic, short-duration disturbances, such as storms, fires, pollution episodes, or harvest removal, because organisms and ecosystem processes are adapted to natural disturbances that occur and have occurred throughout geological and human history. Some organisms actually require periodic disturbance for their long-term persistence, as was noted in our discussion of fire-adapted vegetation. What is new in recent times is the increase in intensity and geographic extent of anthropogenic disturbance, and, of special concern, the large-scale introduction into the environment of new chemical poisons, such as pesticides and radioactive materials. Under the heading "The Poisoning of America," a 1980 news magazine (*Time*, September 22, 1980) reviewed the toxic waste situation as follows:

> Of all of man's interventions in the natural order, none is accelerating quite so alarmingly as creation of chemical compounds. Through their genius, modern alchemists brew as many as 1000 new concoctions each year in the U.S. alone. At last count, nearly

50,000 chemicals were on the market. Many have been an undeniable boon to mankind—but almost 35,000 of these used in the U.S. are classified by the federal EPA as being either definitely or potentially hazardous to human health.

We can add that "potentially hazardous" should include damage to life-support processes, which indirectly damages human health.

While high-level radioactive wastes have been tightly controlled since the beginning of their production, the handling of industrial toxic wastes was until very recently considered a business "externality" not worthy of serious attention. The unwanted material was placed in toxic waste dumps (Figure 16), or just dumped somewhere out of sight, until several local disasters came to public attention. The Love Canal incident in New York State, where a residential development built on top of a waste dump had to be abandoned when people became sick, received wide press coverage. So did the Kepone that poisoned a large section of the James River in Virginia as well as workers in the plant that made the insecticide. (The river recovered but some of the people did not.) In

The Wildlife–Human Connection: Chemically Induced Alteration of Endocrine Function

A large number of human-made, hormonelike, carbon-based chemicals that are being released into the environment (along with a few natural ones) have the potential to disrupt the endocrine systems of animals, including humans. Among these are the persistent, bioaccumulating organohalogens (including some pesticides), many industrial plastics and other products (HCBs, RCBs, dioxins, phenols), and some metals (cadmium, lead, and mercury). Many wildlife populations, including birds, mammals, turtles, and alligators, have already been affected. Abnormal and disruptive behavior and declines in reproductive success have been shown by both laboratory tests and field observations to be linked with body burdens of these chemicals. A decline in human sperm counts and an increase in certain cancers are suspected to be some of their effects on human health. (For more on these alarming trends see Colborn et al. 1996.) As was the case with DDT and CFCs, we can expect much controversy and denial, new regulations, and a frantic search for less deadly substitutes.

(A)

(B)

another well-publicized case, a whole town in Missouri had to be abandoned when material contaminated with the deadly poison dioxin was used to surface streets. Mining and smelting operations, in addition to laying waste to land and soil as noted in the previous section, are a major source of toxic wastes, producing volumes far exceeding that in municipal landfills (Young 1992).

Perhaps the greatest danger and potential disaster is the contamination of groundwater, especially the deep aquifers that provide such a large percentage of the water used by cities, industry, and agriculture (Pye and Patrick 1983). Once polluted, groundwater is difficult, if not impossible, to clean up, since it contains few decomposing microbes and is not exposed to sunlight, strong water flow, or any of the other natural purification processes that cleanse surface water. Already, cities in the industrial heartlands can no longer use local groundwater for drinking because of contamination; they must pipe in water from a distance at great expense—another example of a nonmarket life-support value becoming a cost rather than a benefit to the market economy.

Source Reduction and Cost Internalization

Quite belatedly, these well-publicized incidents and situations involving toxic wastes have created public demand that industry and government work together, first, to clean up the worst dumps, and second, to develop waste reduction and waste detoxification technologies. The next step is to find substitutes for the most toxic chemicals in order to reduce the output of materials requiring special care. Most of all, the cost of waste management needs to be "internalized," that is, to become part of the total cost of production. Once this is done, market pressure will encourage industry to reduce costs by reducing the use of toxic substances that are expensive and dangerous to deal with. In this way the more burdensome regulations can be replaced by economic incentives. We can be optimistic that public opinion will soon insist that the regulatory–incentive infrastructure necessary to promote internalization of costs be put in place.

FIGURE 16
(A) An open garbage dump, no longer permitted in the United States and Canada, but common in many other countries. (Photograph courtesy of Soil Conservation Service.) (B) A toxic waste dump. (Photograph courtesy of Environmental Protection Agency.) These methods of waste disposal can no longer be tolerated. Waste reduction, combined with "high-tech" waste management and recycling, must now receive the highest priority worldwide.

Green Clean

It has been known for some time that many plants are "hyperaccumulators" in that they tolerate, take up, and concentrate up to a hundredfold toxic metals such as lead, nickel, and mercury. It is presumed that the natural selection advantage of this ability is that such concentrations of toxic ions discourage caterpillars or other herbivores from feeding on the plant. In recent years there has been much interest in propagating and bioengineering such plants to help clean up soil and water contaminated by metals and other toxic materials. This newly emerging field is known as "phytoremediation" or "bioremediation," or in the vernacular, "green clean."

The common water weed parrot feather (*Myriophyll*), some kinds of mustard (*Brassica*), and pennycress (*Thlaspi*) are examples of plants that show promise for phytoremediation (Brown 1995). Geneticists have added a bacterial accumulator gene to the small water plant *Arabidopsis* that enhances its ability to volatilize mercury, thereby removing it from the water and dispersing it into the air. For small, highly contaminated areas these technologies can be effective, but they are no substitute for not spilling or dumping toxic materials in the first place.

The Subsidy-Stress Model

Just about everything civilization does has a mixed impact on the natural environment. Humans enrich as well as degrade the environment. Very frequently this is a matter of scale. Human perturbation often enhances ecosystem response at low levels of impact but degrades it at higher levels. Good agricultural or forest husbandry improves our landscape, for example, but large-scale monoculture, supported by heavy applications of fertilizer and pesticides, destabilizes and poisons it (the "too much of a good thing" syndrome). A little bit of heat or CO_2 or phosphate may enhance the productivity of a body of water if these items are limiting under natural conditions, but large amounts of the same inputs may measurably depress basic functions or adversely affect the welfare and survival of particular species.

Such a "subsidy-stress gradient" can be illustrated by the hypothetical "performance curve" shown in Figure 17, in which increasing intensity of

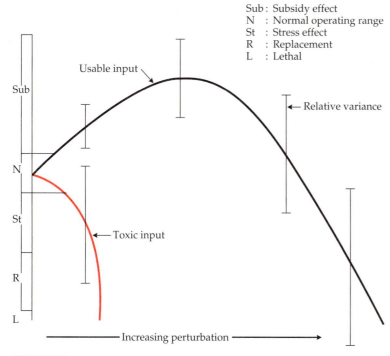

FIGURE 17

Hypothetical performance curve for a perturbed ecosystem subjected to two kinds of inputs. The curves simulate the output response (as measured by appropriate systems or component rates of function) to increasing intensity of input perturbation.

a specific perturbation is plotted on the x-axis and response in terms of some designated rate of function is plotted on the y-axis. If the input perturbation involves usable resources or energy, then a convex or "hump-backed" performance curve may simulate the system response (upper curve, Figure 17). If the perturbation involves toxic substances, or is destructive in some way, then system performance is likely to be reduced—perhaps to a point at which the community is replaced by a new one composed of tolerant species or, ultimately, to a point at which life in the area is severely reduced or exterminated entirely (lower curve, Figure 17).

As the plateau of maximum performance is approached and exceeded, relative variance theoretically increases sharply. Thus, not only the deflection caused by a perturbation is a matter of concern, but also the possibility of reduced stability associated with increased variability. The increase in weather variation resulting from global warming may be an example. (For more on the subsidy-stress model, see Odum et al. 1979.)

Suggested Readings

Abrahamson, W. G. 1984. Fire: Smokey Bear is wrong. *BioScience* 34: 179–180.

Agarwal, A. 1979. Why the world's deserts are still spreading. *Nature* 277: 167–168.

*Altieri, M. A. 1987. *Agroecology: The Scientific Basis of Alternative Agriculture.* Westview Press, Boulder, CO. (Traditional agricultural systems are described on pp. 41–59.)

*Batie, S. S. 1983. *Soil Erosion: Crisis in America's Croplands?* The Conservation Foundation, Washington, D.C.

*Batie, S. S., and R. G. Healy. 1983. The future of American agriculture. *Sci. Am.* 248(2): 45–53.

Berner, E. K., and R. A. Berner. 1987. *The Global Water Cycle.* Prentice-Hall, Englewood Cliffs, NJ.

Blaikie, P. M., and H. Brookfield. 1987. *Land Degradation and Society.* Chapman and Hall, New York.

*Bloom, A. J., F. S. Chapin III, and H. A. Mooney. 1985. Resource limitation in plants—an economic analogy. *Annu. Rev. Ecol. Syst.* 16: 363–392.

Bolin, B. 1970. The carbon cycle. *Sci. Am.* 223(3): 124–132.

Bolin, B., B. R. Doos, J. Jager, and R. A. Warrick, eds. 1986. *The Greenhouse Effect: Climate Change and Ecosystems.* Scope 29, Chichester; John Wiley, New York.

Bormann, F. H. 1985. Air pollution and forests: An ecosystem perspective. *BioScience* 35: 434–441.

Bormann, F. H., and G. E. Likens. 1970. The nutrient cycles of an ecosystem. *Sci. Am.* 223(4): 92–101.

*Brown, K. S. 1995. The green clean. *BioScience* 45: 579–582.

Bryson, R. A. 1974. A perspective on climatic change. *Science* 184: 753–760. (An early warning that atmospheric values such as turbidity and carbon dioxide are being altered by human activities.)

Bullock, T. H. 1955. Compensation for temperature in the metabolism and activity of poikilotherms. *Biol. Rev.* 30: 311–342.

*Carter, V. G., and T. Dale. 1974. *Topsoil and Civilization.* University of Oklahoma Press, Norman.

Chaboussou, F. 1986. How pesticides increase pests. *Ecologist* 16(1): 29–35. (Pesticides sometimes derange plant metabolism, making the plants more vulnerable to disease and infestations.)

*Clark, E. H., J. A. Haverkamp, and W. Chapman. 1985. *Eroding Soils: The Off-Farm Impacts.* The Conservation Foundation, Washington, D.C.

*Clausen, J. C., D. D. Keck, and W. M. Hiesey. 1948. Experimental studies on the nature of species. III. Environmental responses to climatic races of *Achillea.* Carnegie Institution of Washington Publication 581: 1–129.

*Colborn, T., D. Dumandski, and J. P. Meyers. 1996. *Our Stolen Future: Are We Threatening Our Fertility and Survival—A Detective Story.* Dutton (Penguin Books), New York.

*Coleman, D. C., and D. A. Crossley. 1996. *Fundamentals of Soil Ecology*. Academic Press, New York.

*Cooper, C. F. 1961. The ecology of fire. *Sci. Am.* 204(4): 150–160.

*Council on Environmental Quality. 1981. *Environmental Quality*. 12th annual report. U.S. Government Printing Office, Washington, D.C.

Deevey, E. S. 1970. Mineral cycles. *Sci. Am.* 223(3): 148–158.

Delwiche, C. C. 1970. The nitrogen cycle. *Sci. Am.* 223(3): 136–146.

*Ehrlich, P. R., and 19 coauthors. 1983. Long-term biological consequences of nuclear war. *Science* 222: 1293–1300.

*Faulkner, E. H. 1943. *Plowman's Folly*. University of Oklahoma Press, Norman.

Frieden, E. 1972. The chemical elements of life. *Sci. Am.* 227(1): 52–60.

*Gardner, C. 1995. From oasis to mirage: The aquifers that won't replenish. *Worldwatch* 8(3): 30–36.

*Gates, D. G. 1993. *Climate Change and Its Biological Consequences*. Sinauer Associates, Sunderland, MA. (A thorough discussion of the greenhouse effect, climates of the past, and the projected global temperature rise and what it will mean for our ecosystems, particularly in temperate regions of the Northern Hemisphere.)

*Gebhardt, M. R., T. C. Daniel, E. E. Schweizer, and R. R. Allmaras. 1985. Conservation tillage. *Science* 230: 625–630.

*Gliessman, S. R., E. P. Garcia, and A. M. Amador. 1981. The ecological basis for the application of traditional agricultural technology in the management of tropical agroecosystems. *Agroecosystems* 7: 173–185.

*Haagen-Smit, A. J., E. F. Darley, E. F. Zaitlin, M. Hulland, and W. Noble. 1952. Investigation of injury to plants by air pollution in the Los Angeles area. *Plant Physiol.* 27: 18–34.

*Hall, G. F., R. B. Daniels, and J. E. Foss. 1982. Rate of soil formation and renewal in the USA. Chapter 3 in *Determinants of Soil Loss Tolerance*. ASA Special Publication no. 45. American Society of Agronomy: Soil Science Society of America, Madison, WI.

*Hall, J. V., A. M. Winer, M. T. Kleiman, F. W. Lurmann, V. Brajer, and S. D. Colome. 1992. Valuing the health benefits of clean air. *Science* 255: 812–816.

Harris, T. 1991. *Death of a Marsh*. Island Press, Covalo, CO. (Selenium poisoning in the Kesterson National Wildife Refuge and 43 other sites in 15 western states.)

Hobbie, J., J. Cole, J. Dungan, R. A. Houghton, and B. Peterson. 1984. Role of biota in global CO_2 balance: The controversy. *BioScience* 34: 492–498.

Holden, P. W. 1986. *Pesticides and Groundwater Quality*. National Academy Press, Washington, D.C. (Between 1960 and 1980, nitrate concentration increased threefold in Big Spring water that drains 100 square miles of Iowa farmland. Unusually clear-cut case of contamination of groundwater by agricultural chemicals.)

Houghton, R. A. 1987. Terrestrial metabolism and atmospheric CO_2 concentrations. *BioScience* 37: 672–678.

*Hutchinson, G. E. 1944. Nitrogen and the biogeochemistry of the atmosphere. *Am. Sci.* 32: 178–195.

*Hutchinson, G. E. 1950. Survey of contemporary knowledge of biogeochemistry. *Bull. Am. Mus. Nat. Hist.* 95: 554.

Jackson, W. 1987. *Altars of Unhewn Stone*. North Point Press, San Francisco. (Jackson practices as well as writes about ecological agriculture at his Land Institute in Kansas. "Unhewn stone" is an Old Testament metaphor.)

Jenny, H. 1980. *The Soil Resource: Origin and Behavior*. Ecological Studies, vol. 37. Springer-Verlag, New York.

*Jordan, C. F. 1982. Amazon rain forests. *Am. Sci.* 70: 394–401.

*Jordan, C. F. 1985. *Nutrient Cycling in Tropical Forest Ecosystems: Principles and Their Application in Management and Conservation*. Wiley, New York.

Kellogg, W. W., R. D. Cadle, E. R. Allen, A. L. Lazarus, and E. A. Martell. 1972. The sulfur cycle. *Science* 175: 587–596. (Discharges from industrial regions are overwhelming natural removal processes.)

*Kerr, R. A. 1995. Study unveils climate cooling caused by pollutant haze. *Science* 268: 802.

*Kitchell, J. F., R. V. O'Neill, P. Webb, G. W. Gallepp, S. M. Bartell, J. F. Koonee, and B. S. Ausnais. 1979. Consumer regulation of nutrient cycling. *BioScience* 29: 28 -33.

*Kneese, A. V. 1984. *Measuring the Benefits of Clean Air and Water*. Resources for the Future, Washington, D.C.

*Langdale, G. W., A. R. Barnett, R. A. Leonard, and W. E. Fleming. 1979. Reduction of soil erosion by the no-till system in the southern Piedmont. *Trans. Am. Soc. Agric. Engr.* 22: 83–86.

*Langdale, G. W., W. C. Mills, and A. W. Thomas. 1992. Use of conservation tillage to retard erosive effects of large storms. *J. Soil Water Conserv.* 47(3): 157–160. (Areas under conservation tillage suffered no soil loss during heavy rainfall events.)

*Little, C. E. 1987. *Green Fields Forever: The Conservation Tillage Revolution in America*. Island Press, Washington, D.C.

*Lowrance, R., R. Todd, J. Fail, O. Hendrickson, R. Leonard, and L. Asmussen. 1984. Riparian forests as nutrient filters in agricultural watersheds. *BioScience* 34: 374–377.

*Morell, V. 1996. Setting a biological stopwatch. *Science* 271: 905–906.

*Oberle, M. 1969. Forest fires: Suppression policy has its ecological drawbacks. *Science* 165: 568–571. (Fighting fire with fire.)

*Odum, E. P., J. T. Finn, and E. H. Franz. 1979. Perturbation theory and the subsidy-stress gradient. *BioScience* 29: 349–352.

*Opie, J. 1993. *Ogallala Water for a Dry Land*. University of Nebraska Press, Lincoln. (See also *Time*, May 1982, pp. 98–99.)

*Paul, E. A. 1989. Soils as components and controllers of ecosystem processes. Chapter 15 in *Towards a More Exacting Ecology*, ed. P. J. Grubb and J. B. Whittacker, pp. 353–374. Blackwell Scientific Publications, Oxford.

*Phillips, R. E., R. L. Blevins, G. W. Thomas, W. Frye, and S. H. Phillips. 1980. No-tillage agriculture. *Science* 208: 1108–1113. (The authors are the modern-day pioneers in plowless agricultural research.)

Pye, V. I., and R. Patrick. 1983. Ground water contamination in the United States. *Science* 221: 713–718.

*Reich, P. B., and R. G. Amundson. 1985. Ambient levels of ozone reduce net photosynthesis in tree and crop species. *Science* 230: 566–570.

Richards, B. N. 1974. *Introduction to the Soil Ecosystem*. Longman, New York. (Revised in 1987 under the title *The Microbiology of Terrestrial Ecosystems*.)

*Sanford, R. L. 1987. Apogeotropic roots in an Amazon rain forest. *Science* 235: 1062–1064.

Schindler, D. W. 1977. Evolution of phosphorus limitation in lakes. *Science* 195: 260–262. (Nitrogen deficiency can be "self-corrected" by nitrogen fixation, but not so for phosphorus, which is thus the limiting nutrient in the long term.)

*Sidey, H. 1992. Revolution on the farm. *Time* 139(26): 54–56. (The plow is being replaced by new technologies of residue management—no-till, crop rotation, and herbicides that break down quickly—that protect the land and promise even more abundant crops.)

*Soil Science Society of America. 1994. *Defining and Assessing Soil Quality*. SSSA special publication no. 35. Soil Science Society of America, Madison, WI.

*Spencer, D. F., S. B. Alpert, and H. H. Gilman. 1986. Cool water: Demonstration of a clean and efficient new coal technology. *Science* 232: 609–612.

*Steila, D. 1976. *The Geography of Soils*. Prentice-Hall, Englewood Cliffs, NJ.

Svensson, B. H., and R. Soderlund, eds. 1976. *Nitrogen, Phosphorus and Sulfur: Global Cycles*. Ecological Bulletins, 22. Royal Swedish Academy of Sciences, Stockholm.

Turco, R. P. et al. 1983. Nuclear winter: Global consequences of multiple nuclear explosions. *Science* 222: 1283–1292.

United States Department of Agriculture Soil Conservation Services. 1975. *Soil Taxonomy*. Agriculture Handbook no. 436. U.S. Government Printing Office, Washington, D.C.

*Vernadsky, W. I. 1945. The biosphere and the noösphere. *Am. Sci.* 33: 1–12. (Vernadsky's book, *The Biosphere*, was first published in Russian in 1926.)

*Watson, V. J., O. L. Loucks, and W. Wright. 1981. The impact of urbanization on seasonal hydrologic and nutrient budgets of a small North American watershed. *Hydrobiologia* 77: 87–96.

Winfree, A. T. 1987. *The Timing of Biological Clocks*. Scientific American Library, no. 19. Scientific American Books, New York.

Worster, D. 1985. *Rivers of Empire: Water, Aridity, and the Growth of the American West*. Oxford University Press, New York. (A history of the impact of irrigation in arid lands.)

Young, J. E. 1992. *Mining the Earth*. Worldwatch Paper no. 109. Worldwatch Institute, Washington, D.C.

*Indicates references cited in this chapter.

6

Population and Community Ecology

We now come to the more purely biological aspect of ecology, that is, the interaction of organisms with organisms. Up to this point we have focused our attention on the earth's physical and chemical forces. We have outlined how energy from the sun flows through ecosystems such as oceans, croplands, and forests, and we have seen how fuel energy flows guided by humans modify, supplement, and interact with the solar-powered biosphere. Likewise, we have demonstrated how materials are cycled and recycled, and how organisms are limited by temperature, light, nutrients, and other abiotic factors. We have emphasized that organisms are not just pawns in a great chess game in which the physical environment directs all the moves—quite to the contrary, organisms (especially humans) modify, change, and even regulate the physical environment.

With this background, we are now ready to focus attention on the individual, population, and community levels of organization (see Table 1 in Chapter 2). Natural selection and genetic mechanisms operate at these levels to bring about evolutionary changes in species. Understanding how populations grow and how species and individuals react to one another can help us improve the symbiosis between humans and the organisms that provide our life support.

Population Growth Forms

A population, as we defined it in Chapter 2, is a group of organisms of the same species found occupying a given space. Populations have a number of attributes that we are often interested in measuring when we wish to compare different populations, or the same population at different times. Examples include:

Density: Population size in relation to a unit of space.

Natality, or **birth rate**: The rate at which new individuals are added by reproduction.

Mortality, or **death rate**: The rate at which individuals are lost by death.

Dispersal: The rate at which individuals immigrate into the population or emigrate out of the population.

Dispersion: The way in which individuals are distributed in space, generally in one or more of the following three broad patterns: (1) random distribution, in which the probability of an individual occurring in any one spot is the same as the probability of it occurring in any other spot; (2) uniform distribution, in which individuals occur more regularly than at random, like corn in a planted cornfield; (3) clumped distribution (the most common in nature), in which individuals occur more irregularly than at random, as, for example, a clump of plants arising from vegetative reproduction, a flock of birds, or people in a city.

Age distribution: The proportion of individuals of different ages in the population.

Genetic fitness or **persistence**: The probability of individuals' leaving descendants over long periods of time.

The **population growth rate**—the net result of births, deaths, and dispersals—can take a number of forms. Figure 1 shows two contrasting ways that populations grow when the opportunity presents itself, as, for example, at the beginning of a growing season, or when a few individuals enter (or are introduced into) an unoccupied area, or when unused resources become available. For convenience we can label these the **J-shaped** or **exponential growth form** and the **S-shaped** or **sigmoid growth form** (Figure 1). In the former, density increases in exponential or geometric fashion, for example, 2, 4, 8, 16, 32, and so on, until the population runs out of some resource or encounters some other limitation. In theory (Figure 1, upper), growth then comes to an abrupt halt, with density declining rapidly until conditions for another growth period are restored. In practice (Figure 1, lower), the momentum of exponential growth is usu-

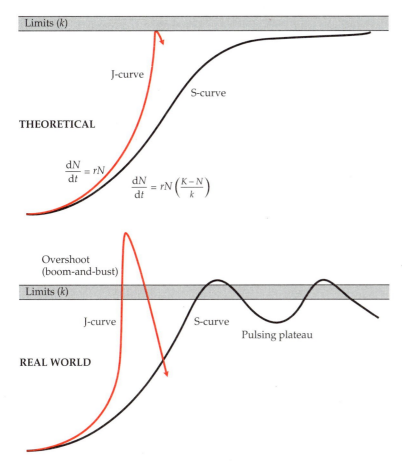

FIGURE 1
Exponential (J-curve) and sigmoid (S-curve) growth in theory and in practice.

ally so great that limits are overshot, resulting in "boom-and-bust" cycles. Populations with this kind of growth pattern fluctuate widely unless regulated by factors outside of the population.

In the S-shaped growth pattern, limiting factors resulting from crowding provide negative feedback that reduces the rate of growth more and more as density increases. Theoretically, if the limitation is linearly proportional to density, the growth form will be a symmetrical sigmoid curve with density leveling off so as to approach an upper asymptote level, K, commonly called the **carrying capacity** because it represents the theoretical maximum sustainable density. In the real world, however, density usually does not level off in a steady state just below K, but pulses above and below the limit because, as discussed in Chapter 3 (Figure 11), there are no set-point controls above the organism level (the pattern exhibits homeorhesis rather than homeostasis). In the absence of outside perturbations, the amplitude of pulses may decrease over time as population and environment adjust to each other.

In Figure 2 the temporarily unlimited J form and the self-limited S form are plotted with density on a logarithmic rather than an arithmetic scale. This type of graph is called a semilog plot since one axis (density in this case) is logarithmic and the other (time in this case) remains arithmetic. Such a plot has the advantage that the J-shaped growth curve becomes a straight line whose slope represents the growth rate constant. The sigmoid growth form becomes a convex curve showing how growth rate decreases with density until it is zero. The slope of the tangent at any

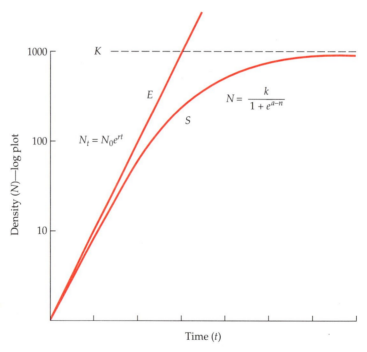

FIGURE 2
Exponential and sigmoid growth plotted on a semilogarithmic graph (density is plotted on a logarithmic scale, time on an arithmetic scale).

point represents the rate of growth at that point in time. As long as growth continues as a straight line on a semilog plot it can be said to be **exponential**. (The differential equations shown in Figures 1 and 2 will be explained below.)

It is evident that exponential growth, whether of a population or of something like society's consumption of fuel, cannot continue for long without the danger of a disastrous overshoot, because with each doubling time the jump becomes larger and larger. If a population of leaf-eating insects in a tree increased exponentially at a rate of tenfold each month, there might be only 100 individuals after two months, but there could be 10,000 after four months, enough to strip the tree bare of leaves. (Such defoliations do occur in nature, as we will see below.) The J-shaped and S-shaped growth forms in Figures 1 and 2 represent models of the extremes of fast and slow growth. Most populations exhibit intermediate growth rates or a combination of patterns. Growth curves for animals with complex life histories, such as insects with larva–pupa–adult stages, often take a stair-step form.

Figure 3 shows three patterns of population growth in relation to population density. Populations that tend to be self-limited in that the rate of growth decreases as density increases can be said to be **inversely density-dependent**. Other populations that tend to grow more or less unrestricted until checked by forces outside the population can be said to be **density-independent**. When poorly constrained by outside factors (e.g., predators, food supply, physical factors), such species are subject to severe oscillations in density and may become serious pests to humans, as already noted. We can perhaps define a "pest" as an opportunist capable of exponential growth when control within the ecosystem is absent or breaks down. We discussed introduced insect and other animal pests as examples in Chapter 2.

There is a third type of relationship between density and growth rate. In some species of social animals and colonial plants, the growth rate is greater at an intermediate density than at either a low or a high density; in other words, both "undercrowding" and "overcrowding" are limiting. Such a pattern is called the **Allee growth form** after the late W. C. Allee, whose 1951 book, *Cooperation among Animals, with Human Implications*, summarized a lifetime of experimental studies of the social life of animals. Species of gulls that nest in colonies are good examples: the number of young produced per pair is higher when the density of the colony is fairly high than when only a few pairs are present or when there is severe crowding. Behavior patterns necessary for pair formation and the care of young are apparently stimulated and augmented by the presence of

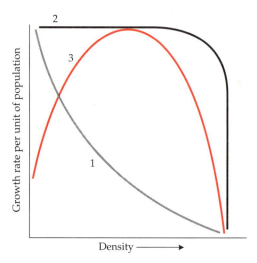

FIGURE 3
Three patterns of population growth rate in relation to population density. Curve 1, growth rate decreases as density increases (self-limiting or inverse density-dependent type); curve 2, growth rate remains high until density becomes high and factors outside of the population become limiting (density-independent type); curve 3, growth rate is highest at intermediate densities (Allee type).

nearby individuals. Oysters too are more successful when the local density on an oyster reef is moderately high, but for a different reason. The oyster begins life as a free-swimming planktonic larva that must settle on a hard substrate to metamorphose into a bivalve adult. Where a large number of old oysters are present, their shells provide a favorable substrate for the larvae to settle upon. If too many oysters are harvested, regrowth of the colony may be slow (or may not occur) for lack of substrate.

Some animal species are famous for periodic **population irruptions** or **outbreaks**, that is, boom-and-bust cycles of abundance that may have a regular, almost predictable, periodicity. Arctic lemmings, which are small, mouselike rodents, irrupt every four years or so. At the peak of their abundance, incredible numbers of lemmings appear on the tundra landscape; one year later only a few can be found. Perhaps the most famous case is that of the snowshoe hare and the Canadian lynx, both of which reach peak abundance at the same time about every 11 years, as indicated by fur-trapping records. Also well documented are the outbreaks of needle-eating caterpillars (the larvae of moths) that occurred

> ## Too Much of a Good Thing?
>
> Allee's principle is relevant to the human condition. Aggregation into towns and urban districts is generally beneficial for our highly social species. But cities, like gull colonies, can become too large and crowded for their own good, and if they do, diminishing returns will set in. For a city, some of these might be increased pollution, noise, crime, and high taxes.

every 5–10 years between 1880 and 1940 in German pure-pine forests. Periodic irruptions taking place over a decade or more occur in populations of spruce budworms, bark beetles, tent caterpillars, and grouse in northern North America. In Eurasia, records of periodic swarms of migratory locusts with numbers capable of wiping out crops in a few hours go back to antiquity (Carpenter 1940).

Irruptions and outbreaks such as these have been most often recorded where the biological community is simplified (where there are few species in each trophic level) either by severe natural limiting factors (as in the Arctic) or by human interference, both of which tend to reduce feedback control mechanisms. Ecologists do not agree on whether this instability is due to external pulses or to biologically mediated factors operating within the populations. As far as tent caterpillars and budworms are concerned, extensive research and experimentation has failed to link any physical environmental factor with the boom-and-bust cycles in northern forests. The latest findings indicate that special viruses are involved in the alternation of strong and weak generations. In a recent review, Meyers (1993) stated that "these insects seem to be adapted by years of evolution to insure that the outbreaks do not devastate the host trees," which recover quickly from the periodic defoliation. If this is true, then spraying with insecticide to "control" the insects is futile once the outbreak is in progress.

r- and K-Selection

Readers who have studied calculus will recognize the differential equations for sigmoid and exponential growth as given in Figure 1, and the integrated forms in Figure 2. The differential forms of these equations can be written in word form as follows:

(1) growth rate = reproductive rate *r* × number *N*

(2) growth rate = reproductive rate *r* × number *N* × self-limiting factor such as (*K* − *N*)/*K*

The two important constants in these equations are *K*, the upper carrying capacity level as already noted, and *r*, which represents the inherent or intrinsic rate of growth of the population when in an unlimited environment. In uncrowded environments (e.g., during early colonization of an abandoned crop field or a new pond) natural selection pressure favors species with high reproductive potentials (large investment in offspring). In contrast, crowded conditions (as in a mature forest) favor organisms with lower growth potential but greater capabilities for utilizing and competing for scarce resources (greater energy investment in the maintenance and survival of the adult). These two modes are known as ***r*-selection** and ***K*-selection** (and species exhibiting them as *r*-and *K*-strategists), based on the *r* and *K* constants in the growth equations.

Table 1 compares the reproductive strategies of two herbaceous plant species. The ragweed, an *r*-strategist, produces numerous seeds and puts 30 percent of its biomass production into reproductive structures (flowers, seeds, etc). In contrast, the forest-dwelling toothwort, a *K*-strategist, produces few seeds and puts only 1 percent of its biomass into reproductive structures; to survive in the heavy shade of the forest, it must put most of its production into nonreproductive structures (stems, roots, and leaves). Of course, we can expect to find reproductive strategies intermediate between *r* and *K* types, as in the case of the six species of goldenrods shown in Figure 13 in Chapter 4.

TABLE 1   Contrasting Reproductive Strategies in Two Herbaceous Plants: Examples of *r*- and *K*-Selection in Species Living in Different Environments

Community and Species	Average No. Seeds per Individual	Percent Dry Wt. in Reproductive Structures
One-year field *Ambrosia artemisifolia* (ragweed)	1190	30
Forest *Dentaria laciniata* (toothwort or pepper-root)	24	1

Data from Newell and Tramer 1978.

Carrying Capacity Revisited

However populations grow, diminishing returns for further growth (dis-economies of scale) set in sooner or later. How, then, do we define and assess the theoretical upper limit to growth (K, the carrying capacity) as the population size that can be sustained in a given environmental situation? The concept of carrying capacity seems straightforward enough, but applying it to humans, and to species with widely different growth forms and life histories, is not easy. Even more difficult is applying the concept to higher levels of organization: populations, communities, and ecosystems. Biologists generally define **carrying capacity** as the number or biomass of organisms that a given habitat can support. Two levels are typically recognized: the **maximum** or **subsistence density**, or the maximum number of individuals that can eke out an existence in the habitat, and the **optimum** or **safe density**, a lower density at which individuals are more secure in terms of food, resistance to predators, and periodic fluctuations in the resource base. In terms of the sigmoid growth form (see Figure 1), K would represent the maximum level, while the optimum level might be somewhere between this saturation level and an **inflection point**, I, where population growth rate is highest and density is one-half to two-thirds as great.

One of the best experimental studies of animal population carrying capacity is that of McCullough (1979), who worked with a deer population enclosed in a two-square-mile fenced area in Michigan. He found that the deer herd "tracks K" in that in the absence of predation, the herd's numbers increase rapidly to maximum carrying capacity and then overshoot it; as a result, the habitat is damaged by overbrowsing and the

Errington's Muskrats

One of the first biologists to write about safe or optimum carrying capacity was Paul Errington (1946a), who is especially well known for his long-term studies of muskrats and mink in Iowa freshwater marshes. He observed that muskrats that had secure dens near feeding areas were much less vulnerable to predation by mink than muskrats that did not have such quality housing. The number of muskrats with safe dens represented the safe carrying capacity, which was almost always smaller than the number that could be supported by the food supply in the absence of predators.

vegetation no longer supports as many deer as before. To avoid this undesirable situation, deer were removed (i.e., experimental predation) at a rate that would maintain the population safely below the overshoot level. On the basis of this long-term study, the maximum carrying capacity was estimated to be 90 deer per square mile, and the optimum carrying capacity to be 50 per square mile.

These concepts, of course, apply equally well to plant and animal populations. For example, 100 pine trees might exist crowded together on a small plot of land, but the individual trees would be stunted in growth and mortality would be high. Reducing density to half by thinning would increase the quality of each individual tree and result in a greater economic value for the stand.

Basing carrying capacity on numbers or biomass is feasible as long as each unit (each individual or unit of weight) has more or less the same impact on the environment as any other unit. However, individuals within a species may differ widely in their impact, as would a butterfly and its caterpillar, for example. This is especially true of humans. As we noted in Chapter 4, per capita consumption of energy and resources in an industrialized country may be 50 times that in a poor country. Accordingly, the carrying capacity, in terms of numbers of persons that can be maintained at their current lifestyle by a given resource-energy base, would be much lower in an industrialized country.

Because humans vary so widely in their impact on life-supporting resources, social scientists add a second dimension, **intensity of use**, to their concept of carrying capacity. For example, William R. Catton (1987) defines carrying capacity as the volume and intensity of use that can be sustained without degrading the environment's future suitability for that

Humans and Carrying Capacity

The message that the optimum carrying capacity is nearly always less than the maximum is a hard one to get across to real estate developers, who have a tendency to want to overstock a neighborhood because there is often quick money to be made by promoting quantity over quality. Strong zoning policies and reductions in tax incentives for development can reduce the human tendency to overshoot.

use. The two dimensions, number of users and intensity of per capita use, will track in a reciprocal manner; that is, as the intensity of use goes up, the number of users that can be sustained goes down, and vice versa. Catton also suggests that carrying capacity can be related to Liebig's law of the minimum (see Chapter 5), which can provide a means for estimating upper limits. From an overall ecological viewpoint, the sociologist's more comprehensive concept of carrying capacity is superior to the biologist's one-dimensional concept, since per capita impact does vary in many species in addition to humans.

It is important to note that carrying capacity is not a constant (as is the K in growth equations) for any given population, but neither is it infinite for the human population. If two populations are competing for the same resource, and one population is removed, the carrying capacity of the other population may be increased. You can think of many other factors that could change the growth plateau.

In theory, the carrying capacity for natural ecosystems at the global level should be nearly constant as long as the amount and quality of solar energy input remains constant. As ecosystems become larger and more complex, the proportion of gross production that must be respired to support the communities increases and the proportion that can be used for further growth in size declines. When these inputs and outputs balance, the size (biomass) cannot increase further without bringing on boom-and-bust cycles.

Optimizing Energy Use

Analyzing how an organism or a population partitions available energy (input) among various activities is analogous to a cost–benefit analysis of one's personal or business finances, with the benefit being increased fitness (survival of the species for now and in the future) and the cost being the energy and time required to insure future reproductive output. Paralleling the partitioning of energy between P (production) and R (respiration or maintenance) and the concept of net energy in the ecosystem as a whole, as discussed in Chapter 4, individual organisms and their populations can grow and reproduce only if they can acquire more energy than is needed for maintenance, a quantity often called **existence energy**. Additional or **net energy** is required for reproduction (for reproductive structures, mating activities, production of seeds, eggs, or young, parental care, and so on) and therefore for the survival of future generations.

Through natural selection, organisms achieve as favorable a benefit–cost ratio of energy to time as possible. For autotrophs, this ratio involves usable light (convertible to food) minus the energy required to maintain the energy-capturing structures (leaves, for example) as a function of the time that light energy is available. For animals, what is critical is the ratio of utilizable energy in food to the energy cost of searching and feeding. Optimization can be achieved in two basic ways: by minimizing time—by efficient searching or conversion, for example; or by maximizing net energy—by selecting large food items or other easily converted energy sources.

Recent energy budget analyses indicate that the lower the absolute abundance of food (or other energy source), the larger the habitat area foraged and the greater the range of food items taken. Experiments conducted by Werner and Hall (1974) provide an example. These investigators presented bluegill sunfish with different sizes and numbers of cladocerans (water fleas) and recorded which sizes of prey were selected. When food abundance was low, prey of all sizes were taken as encountered. When abundance of prey was increased, the fish ignored the smaller sizes and concentrated on the largest cladocerans. The fish thus switched from being "feeding generalists" to being "feeding specialists" as food abundance increased (and vice versa when it declined).

Hunting versus Web-Building Spiders

Spiders are very successful organisms—and interesting to study if one can get over being afraid of them (with few exceptions, they pose no threat to humans and are actually beneficial to us as major predators of insects). Spiders exhibit two different lifestyle strategies: hunters seek out prey on "foot" (Figure 4A), while web-spinners lie in wait for prey (Figure 4B). Since webs have a high protein content, silk formation has a high energy cost, but many spiders recycle the silk by eating it as they repair or rebuild their webs, thus cutting the cost. Peakall and Whit (1976) estimate that the total energy cost of the webs of spiders that recycle silk is less than the energy expended in hunting by some non-web-building species. (This provides a possible lesson for humans: a population that builds expensive labor-saving devices may be able to reduce the costs by recycling the materials.)

Use of Space

Organisms, like people, sometimes aggregate for mutual benefit, and sometimes isolate themselves for individual benefit. Both strategies can have survival value depending on circumstances and species; many species alternate the two modes with the seasons (aggregate in winter,

(A)

(B)

FIGURE 4
Two spiders with different strategies. (A) The wolf spider hunts for prey on "foot." (B) The orb weaver spider captures its prey in a web. (Photographs by P. J. Bryant, University of California, Irvine/Biological Photo Service.)

isolate in summer, for example). Many plants and animals (especially sessile, or permanently attached, ones) form dense colonies (with benefits that we have already noted in our discussion of the Allee effect), while other plants, such as certain desert shrubs, isolate themselves by producing repellent chemicals that discourage close neighbors. By doing so they reserve for themselves scarce soil moisture that might not be sufficient to support more than one individual in a given space. Some highly mobile animals even go so far as to establish property ownership, a behavior pattern known as **territoriality**. Many of our best-loved songbirds do this. A male robin, for example, will stake out an acre or so of optimal habitat (a lawn with trees, for example) at the beginning of the nesting season and defend it against intruders, with the result that no other male robin can enter the defended territory. Much of the loud bird song we hear in the spring is for the purpose of announcing ownership of a territory. People always seem disappointed when they learn that the first robin, or other spring arrival, is singing to establish a territory and not to court the female, who often does not even arrive on the nesting grounds until a week or so after the males. A male that is successful in acquiring and holding a territory has a high probability of attracting a female and breeding, while a male that is unable to establish a good territory will not breed unless he is able to replace a deceased territory holder. In many species, once the breeding pair has formed, the female joins the male in the defense of the territory.

Territorial behavior is most pronounced in vertebrates and in certain arthropods that have complex reproductive behaviors involving parental care. In birds, the defended territory serves the purpose of insuring that the complex business of mating, nest building, and care of eggs and young will not be interfered with by the presence of other birds of the same species. Some ecologists also contend that territoriality has a long-term benefit in keeping population size (density) well within the limits of food supply. We can turn to spiders again for a possible example. In an experimental study of a territorial species of desert spider, Riechert (1981) found that territory size was fixed (only so many spiders could occupy the experimental area), and that this size was adjusted to times of lowest prey abundance. Accordingly, density would not increase beyond an upper limit set by the number of available territories, no matter how much food was available in favorable times—or how much was experimentally introduced into the area. Territory holders occupied the best sites and were much more successful in producing young than the "floaters" (individuals roaming about in unfavorable habitats and unable to establish territories). In this case the potential of territoriality to limit population size and to select the most genetically fit individuals seems to be realized.

Genetic Diversity

Maintaining **genetic diversity** (e.g., heterozygous alleles and balanced polymorphism) is important for the survival of a species. Species can become **endangered** (i.e., in danger of extinction) if population size becomes small and a genetic bottleneck develops (see Chapter 7). Increasing numbers of plant and animal species are endangered or have become extinct due to the destruction of habitat or the breaking up of habitat into isolated patches as a result of human activities. In the United States, passage of a federal Endangered Species Act has resulted in greater effort by both governmental and private agencies (1) to set aside suitable habitat by means of purchase or easement, (2) to maintain or create corridors or strips of habitat connecting isolated patches, which organisms can use as natural highways, thus increasing genetic exchange (Harris 1984); and (3) as a last resort, to introduce into endangered populations individuals from areas where the species is not endangered, or individuals bred or propagated in captivity.

As humans preempt more and more resources and convert more and more natural environment into domesticated habitat, some loss of species is inevitable. However, diversity at all levels (genetic, species, and landscape) can be maintained (we have the "ecological technology") once we understand that it is in our best interest to do so (Myers 1983; Ehrlich and Mooney 1983; Norton 1986; Wilson 1988).

Metapopulation Dynamics

The metapopulation (literally, "among or between populations") is a level between the organism and the species population in the ecological hierarchy (see Table 1, Chapter 2). This level has become a recent focus in conservation ecology because groups of individuals of many species occur in discrete patches or "islands" of suitable habitat separated by unsuitable habitat, but connected by dispersal (Hanski 1989). According to metapopulation theory, the group of individuals in each discrete patch will go extinct at some point in time, but the patch will be recolonized by individuals from a nearby patch. If colonization and extinction balance over a large area or landscape, the total population size may remain about the same. Accordingly, survival of the species may depend more on dispersal (ability to emigrate and immigrate from one patch to another) than on births and deaths within patches. The metapopulation concept, along with the theory of island biogeography to be discussed in the next chapter, provides models not only for the conservation of endangered species but also for wildlife management and pest control (Levins 1969).

Human Population Growth

Our human population has experienced just about every kind of growth form imaginable, including negative growth, as in the fourteenth century when the bubonic plague (the Black Death) reduced the population of Europe by 25 percent (Freedman and Berelson 1974). For many centuries the human population grew slowly, if at all. Then came two periods of more rapid increase tied to energy procurement. The first major increase came with the development of agriculture (which increased the carrying capacity of a given area of land) beginning about 8000 years ago. The second and more rapid increase started about 200 years ago with the Industrial Revolution, the development of the fuel-powered system, the colonization of thinly populated continents, and a decrease in the death rate resulting from advances in medicine and health care. About 80 percent of the increase in human numbers since the dawn of man has occurred during the past two centuries. The rate of increase of the current world population, now over 5 billion, shows signs of slowing, and some demographers (population experts) are bold enough to predict that our numbers will level off (as in the sigmoid growth form) at between 9 and 14 billion sometime in the next century.

Three factors, according to Bongaarts (1994), are expected to determine the size at which the world human population will level off (Figure 5). Even if every woman on earth had only two children that survived to adulthood (the so-called replacement rate) from now on, world population would increase to at least 7 billion because such a large proportion of people in undeveloped countries are just reaching breeding age; this is the "population momentum" factor shown in Figure 5. A second factor, "unwanted fertility" (teenage pregnancies, for example), could raise the

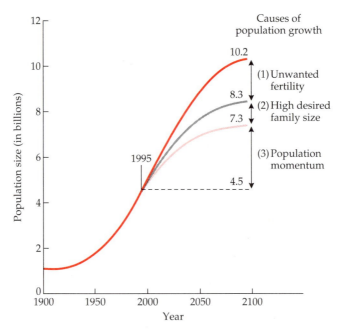

FIGURE 5
Alternative projections for future human population size, showing the effects of three factors that influence human population growth. (After Bongaarts 1994.)

level to at least 9 billion, if left unchecked. A third driving force, "high desired family size" (that is, the perception that children are needed for labor and to take care of parents in their old age) could raise the projected plateau to 10 billion or more.

We—that is, human society—cannot do anything about momentum, but efforts could be made to reduce the other two factors if enough people and political leaders were convinced that such efforts to reduce world population growth would result in a better and safer world. To what extent 10 billion or even 7 billion people would overshoot safe carrying capacity will be discussed in the Epilogue.

A good way to visualize human population growth is in terms of **doubling time**, that is, the number of years required for the density to double. Annual growth is generally expressed in terms of the number of persons added per thousand or per hundred, expressed as a percentage. Thus, a growth rate of 2 percent per annum means that 2 persons per 100, or 20 per 1000, are added each year. Since over the years the people added to the population produce more people, growth at 2 percent is more than just adding 2 per 100 each year; the population actually increases in compound interest fashion, as does your money in a savings account. An approximate estimate of the doubling time, t, for a 2 percent rate, r (expressed as a decimal fraction), can be obtained as follows:

$t = \log 2/r$

$t = 0.6931/0.02 = 35$ **years**

Two percent does not seem like much, but at that rate, the population of your city would double twice in a person's lifetime. At present, the world population is growing at a rate of about 1.8 percent. In many undeveloped countries the rate is 3 percent (doubling time about 23 years) or more, but in developed countries the rate is dropping below 1 percent for the most part (doubling time 70 years or more), with some European countries exhibiting zero growth. Brown and Jacobson (1986) show that the world is sharply divided between countries with low growth rates (average 0.8 percent) and those with high rates (average 2.5 percent). Urban growth, however, is higher than 3 percent in many parts of the world (rich and poor countries alike) as people swarm from the rural countryside to the metropolitan districts seeking a better economic life. One forecast is that 80 percent of the world's people will be urban by the turn of the next century. The problem is acute not only in poor countries but also in rich countries, where immigration from the rest of the overcrowded world is becoming a major concern.

Many sociologists and economists believe in, or at least have high hopes for, what is known as the **demographic transition**, the idea that human population growth slows as people become more affluent and less dependent on their children for labor. With increasing affluence, people do tend to have fewer children and to put more of their energy and money resources into improving their own quality of life—a shift to a *K*-type lifestyle. According to this view, the population problem is an economic problem, but much controversy remains (Teitelbaum 1975; McNa-

A Tale of Two Islands

In 1944, 29 reindeer (24 females and 5 males) were introduced to the uninhabited 128-square-mile island of St. Matthew in the Bering Sea, midway between Alaska and Russia. In the absence of predators or any other factor that might limit population growth, the herd increased rapidly to about 6000 in 20 years, well above, as it turned out, the carrying capacity of the thin tundra vegetation. With the lichens, the most nutritious available food, grazed almost to extinction, the herd was in such poor condition that all but a few females died during the heavy snows of 1963–1964, in a spectacular boom-and-bust cycle with no recovery (Klein 1968).

Incredible as it might seem, the same boom-and-bust cycle happened to the human population of Easter Island in the South Pacific, only it took centuries rather than years. Compared with St. Matthew, Easter was a tropical paradise, with a mild climate and luxurious forests, when it was colonized by a few boatloads of Polynesians some 1600 years ago. The colonizers took dominion over the land, multiplied, and prospered until their numbers and resource consumption became too great, whereupon they fragmented into warring tribes that began killing each other off (Catton 1995). Today, all that is left is a barren grassland and the strange stone statues that stand in silent testimony to a culture that consumed its island life-support systems and then consumed itself.

One would think that humans would be smarter than reindeer and recognize the limitations of islands, with their sharp boundaries and obviously limited space. But the human mindset seems to assume that emigration and outside trade can provide the "antacid medication" for overindulgence. Many island states today are indeed increasingly dependent on tourism and imports as antidotes for too much growth (Acharya 1995).

Egypt: A Malthusian Microcosm

The English clergyman Thomas Malthus is well known for his 1798 essay in which he predicted that humans would increase faster than their food supply, thereby condemning humankind to life on the edge of starvation. So far, this has not happened worldwide, but the current situation in Egypt shows that human numbers can in a very short time outpace the food supply.

Between 1900 and 1990, arable land in Egypt increased from 5.5 to 7.5 million acres as the result of the Aswan Dam and irrigation of dry lands, but during the same period its population increased from 10 million to 50 million. Accordingly, per capita arable land decreased from 0.5 to 0.1 acres. Despite increased yield (more food per acre), Egypt has to import food to provide its people with even the barest essential diet (Biswas 1993).

Egypt seems to be a case in which the promise of technology (dams, high-yielding cultivars, etc.) simply discouraged any attempt to limit population growth—that is, until very recently when, quite appropriately, the first international conference on human population growth was held in Cairo!

mara 1982). The key question is: Does human population growth spur economic development, or restrain it? Two National Academy of Sciences reports published 15 years apart (1971 and 1986) agree that rapid population growth has no economic or other benefit because social and environmental problems are created faster than they can be solved. Accordingly, slower growth than is now occurring would be beneficial to economic development and quality of life for individuals in most countries of the world. Brown and Jacobson (1986), Abernathy (1993), and others are skeptical that the demographic transition is working or will work in countries with high growth rates. Most demographers believe these countries must take political action to control population growth, as China is now doing.

Interactions between Two Species

The effect that one species may have on the population growth and well-being of another species may be negative (−), positive (+), or neutral (0).

Population Control in the Harvester Ant

In a study of harvester ants (so called because they collect and store seeds in underground galleries), Gordon (1995) reported that colonies grow in sigmoid fashion, leveling off in density after about 6 years. As the number of ants increases and more and more ants return to the colony without seeds, the rate of "antennal contacts" (which are the way ants communicate with each other) rises, and fewer eggs and larvae are produced. Unfortunately for humans, increased interaction between people in crowded cities seems to increase violence, but does not decrease birth rates.

Theoretically, then, populations of two species may interact in basic ways that correspond to the nine possible combinations of 0, +, and −, namely, 00, − −, ++, +0, 0+, 0−, −0, +−, and −+. A coordinate model of the most important of these interactions is shown in Figure 6. The terms that are used to designate these types of interactions are as follows:

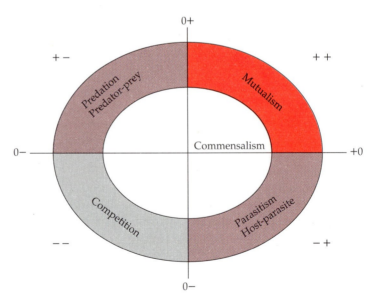

FIGURE 6
Coordinate model of two-species interactions showing the five most common relationships.

Competition (– –): both populations inhibit or have some kind of negative effect on each other
Predation (+ –): positive for the predator, negative for the prey
Parasitism (– +): negative for the host, positive for the parasite
Commensalism (+0): one species, the commensal, benefits, the other is
  not affected
Cooperation or **mutualism** (+ +): both populations benefit from the interaction, which may be optional (cooperation) or essential for the survival of both partners (mutualism)

Competition

The word *competition* denotes a striving for the same thing, as in the familiar case of two businesses striving for the same market. At the ecological
level, competition becomes important when two organisms strive for
something that is not in adequate supply for both of them. Thus, plants
compete for light and nutrients in a forest, and animals compete for food
and shelter when these resources are relatively scarce in terms of demand.
Competition can also take the form of mutual inhibition when two organisms directly interfere with each other while striving for something, even
though that something is not in short supply. For example, the organisms
might excrete substances that interfere with each other, or they might eat
each other.

The result of competition is that both parties (that is, the competitors)
are hampered in some manner (thus, the – – designation). At the population level, this means that density or rate of population growth will be
reduced or held in check by the competitive action. Both competition
within species (**intraspecific**) and between two or more species (**interspecific**) are important in determining the number and kinds of organisms to
be found in a given habitat or community. Intraspecific competition is an
important factor in those populations that tend to be self-regulated, as in
the sigmoid growth form, and in those that are territorial.

As might be expected, interspecific competition is most pronounced
between species with similar lifestyles and resource needs. In natural
communities, closely related organisms having similar habits generally do
not occur in the same place. If they do, they often use different resources
or are active at different times. The explanation for ecological separation
of closely related (or otherwise similar) species has come to be known as
Gause's principle, after the Russian biologist G. F. Gause, who in 1932 first
observed such separation in experimental cultures of protozoa. Later
Hardin (1960) suggested **competitive exclusion** as a more descriptive term
for this principle.

Competitive Exclusion versus Coexistence

As in human affairs, competition in the world of nature has its overall beneficial as well as its negative aspects. The more fit and better adapted replace the less successful. Diversity and adaptiveness are enhanced as species strive to avoid the negative effects of competition by seeking new habitats or resources. If competition between two species is severe, one may be eliminated entirely, or forced to occupy another space or use a different food or other resource. Or the two species may be able to live together at reduced densities by sharing the resource in some sort of equilibrium. These two possibilities—competitive exclusion and coexistence—are shown in Figure 7 in terms of growth form models based on two experimental studies. One study (Figures 7A and 7B) involves two closely related species of animals (beetles), and the other (Figure 7C) two species of plants (clovers).

Thomas Park and his students and associates at the University of Chicago carried out a long series of competition experiments with laboratory cultures of flour beetles in the 1940s and 1950s. These small beetles (several species in the genus *Tribolium*) are major pests in stored food, but they do have a redeeming feature as useful laboratory animals. The beetles complete their life cycle in a container of flour or wheat bran; the medium is both food and habitat. If fresh medium is added at regular intervals, populations of beetles can be maintained indefinitely. Park's experimental setup can be thought of as a stabilized heterotrophic micro-ecosystem, or **microcosm**, in which imports of food energy balance heat and respiratory losses. Thus, the microcosm resembles on a small scale the city or oyster reef discussed in Chapter 3 (see Figure 5 in Chapter 3).

The Chicago investigators found that when two different species of *Tribolium* are placed in this little homogeneous universe, invariably one is eliminated sooner or later while the other continues to thrive. In other words, one species always wins the competition. Two species of *Tribolium* cannot both survive when confined within the same jar, even though either one does quite well when alone in the culture—a clear case of competitive exclusion. The relative number of individuals of each species originally placed in the culture does not affect the eventual outcome, but the climate imposed on the microcosm does have a pronounced effect on which species wins out. As shown in Figures 7A and 7B, one species wins when conditions are hot and wet, while the other species survives when conditions are cool and dry. Under intermediate conditions, sometimes one, sometimes the other wins, following the gradient between the extreme conditions; halfway between the extremes either species has a 50–50 chance of surviving.

(A)

(B)

(C)

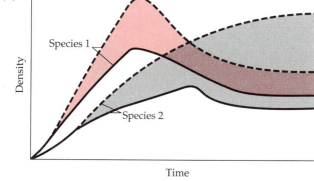

FIGURE 7

Competitive exclusion and coexistence. (A) Flour beetle species 1, *Tribolium castaneum*, excludes species 2, *T. confusum*, in mixed culture when the climate is hot and wet (34°C, 70 percent relative humidity), even though each species does well under these conditions in the absence of the other. (B) Species 2, *T. confusum*, excludes species 1, *T. castaneum*, in mixed culture when the climate is cool and dry (24°C, 30 percent relative humidity), even though both species are able to live separately under these conditions also. (Data from Park 1954.) (C) Two species of clovers, *Trifolium repens* (species 1) and *T. fragiferum* (species 2), are able to coexist in mixed culture, but at a lower leaf density for each species than in a pure culture. (Data from Harper and Clatworthy 1963.)

Experiments with clovers performed by John L. Harper and his associates at the University College of North Wales illustrate that closely related species are able to coexist despite crowding and competition for limited resources. The results of one series of experiments are shown in Figure 7C. Two species of clovers (of the genus *Trifolium*) were able to complete their life cycle and produce seed when thickly planted together in trays, but the density of each species was less than in pure cultures. Small, but important, differences in growth form enable these two competitors to coexist. Species 1, *T. repens*, grows faster and reaches a peak in leaf density sooner, but species 2, *T. fragiferum*, has longer petioles (leaf stems) and higher leaves; it is thus able to overtop the faster-growing species and avoid being shaded out of existence. Harper (1961) concluded that two species can persist together if there are differences in the timing of their growth, in their nutritional requirements, or in their sensitivity to grazing, toxins, light, water, and so on. Also, the coexistence of competitors is enhanced by periodic environmental changes or disturbances that favor first one and then the other species. For example, alternating cool–wet and hot–dry seasons might allow the two species of flour beetles to coexist.

The study of laboratory or greenhouse populations contributes to our understanding of the mechanisms, ecological and genetic, that might operate when species interact. This is evident from the examples given, but, as emphasized in Chapter 2, a multilevel approach is ultimately necessary, since study at each level of organization contributes something, but cannot reveal the total picture. When we move from the study of discrete populations under controlled conditions to the real world of communities and ecosystems, we do indeed find evidence for competitive exclusion. More often, however, we find that species, even those unable to live together in a restricted microcosm, adapt to coexistence by reducing competition pressure through physiological, behavioral, or genetic changes, by exploiting the same resource at different times or places, or by shifting their positions in environmental gradients. In fact, one reviewer of competition theory, den Boer (1986), concludes that coexistence is the "rule" and complete competitive exclusion the "exception" in the open systems of nature.

A classic field study of competition is that of Joseph Connell (1961), carried out on a rocky seashore. In the intertidal zone, sessile animals and plants such as barnacles, mussels, oysters, and seaweed often occur in bands. In Connell's study area, a small species of barnacle of the genus *Chthamalus* occupied a band near the upper part of the intertidal zone, and a larger species of the genus *Balanus* occupied a wide band below the *Chthamalus* population. Such a zonal distribution strongly suggests competitive exclusion, but since there are other possible explanations for the distribution, removal experiments were required to test this hypothesis. When Connell removed the large species (*Balanus*) and kept new individuals from settling, the small species (*Chthamalus*) invaded the upper part of the *Balanus* zone and grew well where it was not normally found. However, when *Chthamalus* was cleared away, *Balanus* did not extend into the vacated territory; its larvae were unable to survive in the more exposed upper zone even in the absence of a potential competitor, as was determined by subsequent observation and experimentation. This kind of interplay between physical limitations and competition has proven to be a common pattern in many environmental gradients.

Predators and Prey

To a great many people, the word *predator* brings to mind a fierce, cruel animal that we could well do without. In truth, however, predation plays an important role in the economy of nature, and is beneficial to the human economy as well when it comes to the control of insects and other pests. Our aversion to predators is paradoxical, since over the centuries humans have been the most destructive predators that ever roamed the earth. To be objective, it helps to think about predation from the population rather than the individual standpoint. Predators, of course, are not beneficial to the individuals they kill, but they may benefit the prey population as a whole by removing unfit individuals and preventing overpopulation. For example, in the absence of predators many species of deer tend to increase in numbers to the point of exceeding their food supply, as was documented in our discussion of carrying capacity earlier in this chapter. In this case, the predator improves the "quality of life" of the prey population.

On the other hand, a predator can be strongly limiting to the point of reducing the prey population to extinction or near extinction. Which situation exists for any pair of interacting species depends on the *degree of vulnerability of the prey to the predator*. From the predator's viewpoint, this depends on how much energy it must expend to capture the prey; from the prey's standpoint, it depends on how successfully individuals are able to avoid being caught and how well they are able to hide or protect their offspring. Prey vulnerability is often increased by human disturbance in the landscape, and especially by the introduction of new predators to which the prey are not adapted. For example, many years ago the mongoose, a weasel-like predator, was introduced on many Caribbean islands in the hope of controlling rats in the sugarcane fields. Instead, it not only did not eliminate the rats, but played havoc with vulnerable ground-nesting birds, reptiles, and turtles (Seaman 1952).

An act of predation, such as a hawk catching a game bird, may be spectacular and easily observed, but many other factors that may be more limiting to prey populations are not as well understood by untrained individuals. The late Herbert L. Stoddard (1936) and his associates, working on Georgia game preserves, showed that hawks are not a limiting factor to quail populations as long as vegetative escape cover lies near feeding areas so that healthy quail can easily escape from attacking hawks (see also Errington 1946b). High densities of quail can be maintained by patch burning and other land management procedures that build up food supply and refuge cover. In other words, when efforts are directed toward improving quail habitat, removing hawks to protect the quail is unnecessary—and even undesirable, because hawks also prey on rodents that eat quail eggs. But "ecosystem management" is more difficult and less dramatic than shooting hawks, and game managers are often pressured by hunters into the latter, even when they know better.

Predators not only have both negative and positive effects on their prey, but also influence the composition of the whole community. For example, rocky intertidal zones, where the substratum available for colonization is limited, are often dominated by one species of shellfish or barnacle. Removal of many individuals of the dominant species by predators (starfish, for example) opens up the habitat for other species, thus increasing the species diversity of the community (Paine 1966). One should not conclude from this one well-studied example, however, that predation always increases species diversity, because it may not have such an effect where habitat is more extensive, where the predator does not have such "easy pickings," or where storms or other periodic disturbances reduce dominance.

Herbivory

When the "prey" is a plant and the "predator" is a plant-eating animal (herbivore or grazer), their interaction is called **herbivory**. Plants cannot escape a would-be consumer by running away or hiding, but they can and do defend themselves by producing antiherbivore structures such as thorns, chemical repellents such as tannins, and substances that are poisonous to animals. In the tropics, where the climate is especially favorable for insects year round, trees employ a double defense: tough leaves (hard, waxy cuticles) and chemical deterrents such as phenols (Coley et al. 1983). The study of plant–herbivore relationships is becoming a "growth industry" in ecology and biochemistry because of the incredible array of complex chemicals that plants produce to defend themselves against insect and other herbivores, which in turn evolve numerous ways to

Natural Insecticides

Selecting or engineering human food plants to produce systemic insecticides could be a good agricultural strategy, but it has drawbacks. The energy a plant puts into synthesizing its chemical defense cuts down its net production (yield), and the antiherbivore chemicals it produces may also be poisonous to humans, or may affect the taste or palatability of the food. But the artificial insecticides we now use in such huge quantities can be even more poisonous. Since natural and artificial insecticides are often similar in chemical composition, we can expect that insects will evolve resistance to both. Thus, control of insect pests in agriculture will be a never-ending battle. However, once we realize that yield is not the only criterion for success in agriculture, then the genetic engineering of plants that can defend themselves against herbivores becomes a worthwhile goal.

detoxify or sequester the repellents. More about this coevolutionary "arms race" will be found in Chapter 7.

Parasites and Hosts

Much of what has been said about predation also holds for parasitism. In fact, parasites and predators form a more or less continuous gradient from tiny bacteria and viruses that live inside the host's tissue to large carnivores that roam the ecosystem at large. The term **parasite** is usually used for a small organism that actually lives in or on a host, which serves as both energy source and habitat. In contrast, we think of a predator as being free-living and larger than its prey, which serves as an energy source but not as habitat. But all kinds of intermediate situations exist.

Although parasitism and predation are similar in terms of ecological interactions, important differences are found in the extremes of each situation. Parasitic organisms generally have higher reproductive rates and exhibit a greater host specificity than do most predators. Furthermore, they are often more specialized in structure, metabolism, and life history, as is necessitated by their special environment and the problem of dispersal from one host individual to another. Some entire classes and orders of organisms, such as the Cestoda among the flatworms and the Sporozoa among the protozoa, have become adapted to parasitism. The most specialized species have very complex life cycles involving a succession of host tissues and even an alternation of host species, as in the case of *Plasmodium*, the parasites that cause malaria in humans and other animals. *Plasmodium* species alternate between mosquitoes and vertebrates during different parts of their life cycle, and each *Plasmodium* species parasitizes a different vertebrate host.

Host specificity in parasites is a very important consideration. Because many species of parasites can live in only one or a few species of hosts, the host–parasite interaction is especially intimate and potentially limiting to both populations. Humans have often been able to utilize parasites to control pests. Insect pests that have been introduced from other parts of the world have often been brought under control by introducing the native parasites that regulated the insect in its original habitat; in other cases artificial propagation of parasites has helped. Practical biological control of this kind is feasible with parasites that are specific for the species one wishes to control. Such a parasite keeps constantly at work and can quickly adjust to increases and decreases in host numbers. In contrast, a pest usually cannot be controlled by the introduction of a generalized parasite or predator, which may itself become a pest if it spreads its attack to species other than the intended target—as in the case of the mongoose cited above. Likewise, we are learning from nature the wisdom of developing species-specific chemical pesticides to replace the broad-spectrum poisons that kill useful organisms along with those judged to be harmful.

An important principle or generality about parasite–host and predator–prey interactions may be stated as follows: The limiting effects of parasitism and predation tend to be reduced and their regulating effects enhanced where the interacting populations have had a common evolutionary history in an ecosystem that is stable enough, or spatially diverse enough, to allow reciprocal adaptation. In other words, natural selection tends to reduce the detrimental effects on both interacting populations, since severe depression of host or prey populations by a parasite or predator can only result in the extinction of one or both populations. Consequently, violent parasite–host or predator–prey interactions happen most frequently where the interaction is of recent origin, or where there has been a recent large-scale disturbance, as might be produced by humans or by climatic change.

As noted in Chapter 2 and again in Chapter 3, a list of the diseases, parasites, and insect pests that have caused the greatest losses in agriculture and forestry would include many species recently introduced into a new area or to a new vulnerable host. The European corn earworm, the gypsy moth, the Japanese beetle, and the Mediterranean fruit fly are examples of serious pests introduced into the United States from other continents. The ever-present danger from introduced species is the reason you are severely restricted by customs regulations on what you can take into or out of the country. Much the same principle applies to human diseases: the most feared are the newly acquired. In contrast, where parasites and predators have long been associated with their respective hosts and prey, their interaction is moderate, and is neutral or beneficial from the long-term viewpoint.

A laboratory demonstration of the evolution of reciprocal adaptation in a host–parasite system is illustrated in Figure 8. House flies (*Musca domestica*) and parasitic wasps (*Nasonia vitripennis*) were placed in a multicell cage consisting of 30 plastic boxes, designed to allow some of the flies to escape the wasps and slow down parasite dispersal. When wild stocks were brought together for the first time, populations of both species oscillated wildly. After two years and many generations, the flies developed resistance and the parasites became less active in searching for prey, resulting in a more or less stable equilibrium. The experiment demonstrated how genetic feedback can function as both a regulatory and a stabilizing mechanism in population systems (Pimentel 1968).

In the natural world, more time may be required for adaptation, and there is always the chance of extinction before adaptation can occur. The chestnut blight is a case in which adaptation or extinction has been hanging in the balance for nearly a century. In 1904 a parasitic fungus that attacks the bark and stems of the chestnut tree was accidentally introduced to North America with the importation from China of the oriental chestnut, which is resistant to the disease. The American chestnut, which constituted 40 percent of the biomass in the forests of the southern Appalachians, proved to be extremely vulnerable to the introduced parasite. By 1952, all the large trees had been killed, their gaunt gray trunks becoming a characteristic feature of the Appalachian landscape. The chestnuts continue to sprout from the roots, and such sprouts may produce fruits before they die, but no one can predict whether the ultimate outcome will be extinction or adaptation (Anagnostakis 1982). For now, the species has been replaced by other hardwoods (oaks, chiefly). The forest as a whole has adapted to the loss, since the total biomass of the present-day forest is now similar to that under pre-blight conditions. This case also illustrates how redundancy—that is, the presence of more than one species in a basic functional niche—enhances resilience and recovery at the ecosystem level (see the discussion of biodiversity in Chapter 3).

A good example of reciprocal adaptation involves the myxoma virus and the European rabbit. Rabbits were introduced into Victoria, Australia,

(A) Newly associated

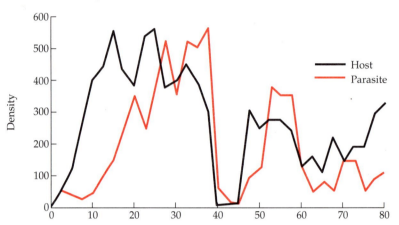

(B) After two years of association

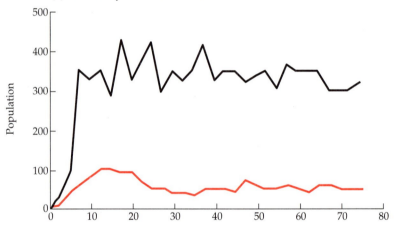

FIGURE 8

Evolution of coexistence in the host–parasite relationship between a house fly (*Musca domestica*) and a parasitic wasp (*Nasonia vitripennis*). Density figures are number of insects per unit in a 30-unit cage. (A) Newly associated populations oscillated violently, as first the host and then the parasite density increased and crashed. (B) Populations derived from colonies in which the two species had been associated for two years coexisted in a more stable equilibrium without crashes. That adaptive resistance had evolved in the host was indicated by the fact that the birth rate of the parasite was greatly reduced (46 progeny per female, compared with 133 in the newly associated system), and the parasite population leveled off at a low density. (After Pimentel and Stone 1968.)

in 1859. They spread rapidly and overgrazed choice rangeland, competing with sheep. The myxoma virus was introduced into Australia in 1950 to control the rabbits. This parasitic virus, which is transmitted to rabbits by mosquitoes, causes myxomatosis, a disease known to control rabbit populations in Europe. As described by Levin and Pimentel (1981), the parasite when first introduced was extremely virulent and killed its host in a few days. However, the first epidemic of myxomatosis killed only 99.8 percent of the rabbit population. Subsequently, the virulent strain was gradually replaced by a less virulent one that took a longer time to kill the host, giving the mosquitoes that transmit the virus more time to feed on infected hosts. Because the avirulent strain did not destroy its food resource (the rabbit) as rapidly as the virulent strain, more and more avirulent-type parasites were produced and were available for transmission to new hosts. Thus, natural selection seems to favor the avirulent over the virulent; otherwise, both parasite and host would eventually become extinct (Alexander 1981; Anderson and May 1982).

As with all generalities, one can expect to find exceptions. In a review of host–parasite relations, Ewald (1983) cautions that the interaction may have various outcomes, including increased disease severity over time, which he believes is the case with human malaria. There is no question that newly acquired human diseases, such as AIDS, and mutated, resurgent infectious diseases, once thought to have been eradicated, do increase in virulence, at least until immune systems can adjust. Platt (1996) lists some of the disruptions that make this adjustment difficult, namely, overcrowding due to overpopulation, lack of clean water and sanitation, poorly planned development, overuse of antibiotics, and, especially, increases in human mobility (such as air travel).

Recent experience in genetic engineering to control pests has demonstrated that control lasts longer if the engineered parasite is not so virulent as to hasten the development of resistance by the pest. A dramatic example is the development of a rust-resistant strain of wheat, accomplished by introducing a "slow-rust" gene that keeps the disease at a low level so that the rust fungus does not have an incentive to mutate (*Science* 258: 551, 1992). The same principle seems to apply to antibiotic medications: overuse simply encourages the disease organism to mutate, creating a new organism that is resistant to the antibiotic. In the same vein, overuse of pesticides helps to produce resistant pests.

Commensalism, Cooperation, and Mutualism

We come now to what might be called the "positive interactions." Charles Darwin's emphasis on "survival of the fittest" directed special attention to competition, predation, and other "negative" interactions. However, cooperation for mutual benefit is widespread in nature and is also important in natural selection.

Several decades after Darwin, in 1902, the Russian Peter Kropotkin published a book entitled *Mutual Aid: A Factor of Evolution*. Kropotkin chided Darwin for his overemphasis on natural selection as a bloody battle (Tennyson's "red in tooth and claw" metaphor). He outlined in considerable detail how survival was often enhanced by, or even dependent on, one individual helping another, or one species aiding another for mutual benefit.

Kropotkin's writings were much influenced by his personal philosophy of peaceful coexistence. Like Gandhi and Martin Luther King, who came along later, he was a firm believer in nonviolent solutions to human conflicts. At the time he wrote his book he was a political refugee living in England. A good deal of his book is devoted to documenting the importance of cooperation in primitive societies, rural villages, and labor union guilds as well as among animals. (For more on Kropotkin, see Gould 1988 and Todes 1989.)

Positive interactions between two or more species take three forms that may represent an evolutionary series. **Commensalism** (+0) is a simple type of positive interaction in which one species benefits and the other is not affected to any significant degree. If two species benefit each other but are not essential for each other's survival, the relationship is usually called **cooperation** (++). If the relationship is so intimate as to be vital or necessary for the survival of both species, we call the interaction **mutualism** (++, sometimes written ++/++ to emphasize the obligatory nature of the relationship) (Dindal 1975).

Commensal relationships are especially common between small motile organisms and larger sessile ones. The edge of the sea is a good place to observe such a relationship. Practically every worm burrow, shellfish, and sponge contains various "uninvited guests" (e.g., crustaceans, sea worms, small fish) that need the shelter or unused food of the host, but do neither good nor harm to the host. If you like oysters, and open up a lot of them, you may find a small, delicate crab living in the mantle cavity of an oyster. These crabs are usually commensal, although sometimes they overdo their guest status and partake of the host's tissues (Christensen and McDermott 1958). It is but a small step from commensalism to parasitism on the one hand or helping behavior on the other. Commensals are not as host-specific as parasites, but some are found associated with only one species of host.

Mutualism, the ultimate in cooperation, is extremely widespread and important. Many pairs or groups of species live together for mutual benefit as obligatory partners (neither can live alone). Indirectly, many mutualisms also benefit the ecosystem as a whole. For example, as described in Chapter 5, the mutualism between plants and nitrogen-fixing microorganisms not only benefits the partners, but also plays a key role in the life-supporting nitrogen cycle.

Mutualism most often involves two species that are different taxonomically (not even distant "cousins"), each of which has vital "goods or services" that the other needs. Microorganisms that digest cellulose and other resistant plant residues are often mutualistic with animals that do not have the necessary enzymes for this. Two examples, ruminants and their rumen bacteria and termites and their intestinal flagellates, were mentioned in Chapter 4 in connection with the detritus food chain. In both cases the microorganisms break down cellulose into fats and carbohydrates the animals can use, and the hosts provide the microorganisms with a place to live and protection against competitors or predators.

An even more intricate interdependence may develop in which the microorganism partner lives outside the body of the host. Tropical leaf-cutting ants cultivate fungal gardens on leaves they harvest and store in their underground nests. The ants fertilize (with their excretions), tend, and harvest the fungal crop much in the same manner as a human mushroom grower. A lot of "ant energy" is required to supply and maintain this monoculture, just as energy is required in an intensive crop culture by humans.

Ants and acacia trees (those picturesque legume trees of tropical savannas) are involved in another striking mutualism. In Africa, the trees house and feed the ants, which nest in special cavities in the branches. In turn, the ants protect the tree from would-be herbivorous insects. When ants are experimentally removed (as by poisoning with an insecticide), the tree is quickly attacked and often killed by defoliating insects. In some areas of the New World tropics, acacias do not house ants but instead protect themselves by producing antiherbivore chemicals. No one has determined which strategy (maintaining an army of ants or producing defensive chemicals) is the most energy-efficient.

Another class of mutualism involves an autotroph—an organism that can make food—and a heterotroph, which cannot but which may be able to provide protection or nutrients for the autotroph. Prime examples are corals, described in Chapter 8, and mycorrhizae, mentioned in Chapter 2 as an "emergent property" and again in Chapter 5 in the discussion of nutrient recycling. Like nitrogen-fixing bacteria with legumes, mycorrhizal fungi interact with root tissue to form composite "organs" that enhance the ability of a plant to extract minerals from the soil. Thin fungal filaments called **hyphae** grow out from the combined fungal–root tissue and are able to extract phosphorus and other scarce nutrients (by chelation or other means not well understood) that would not be available to nonmycorrhizal roots. In return, of course, the fungus is supplied with some of the plant's photosynthate.

There are two main types of mycorrhizae, shown in Figure 9. In the **ectomycorrhizae**, the fungus forms a sheath or network around actively growing roots from which hyphae grow out into the soil, often for long distances. These associate mostly with trees, especially pines and other conifers and tropical trees. The **vesicular–arbuscular** or **VA mycorrhizae** (formerly called endomycorrhizae) penetrate into root tissue where they form characteristic vesicle-like structures (hence the name). Hyphae extend out into the soil as in the ectomycorrhizae. These colonize all but a few genera of plants of all forms, including herbs, crops, shrubs, and trees in all climatic regions.

Mycorrhizae are not generally host-specific, which means that they can often colonize whatever plant root comes in contact with their spores. Some ecto types produce large aboveground sporocarps or mushrooms that facilitate dispersal. The VA types produce spores underground, where they may be dispersed by soil-dwelling animals. (For more on mycorrhizae see Wilde 1968 and Allen 1991.)

The role of mycorrhizae in direct mineral cycling and their importance in the tropics and in crop production was emphasized in Chapter 5. It is fortunate that the pine–mycorrhizal mutualism does so well on the millions of acres of the southern United States where the topsoil was eroded away under the row crop–tenant farming system that persisted for too long. Otherwise, many of these eroded acres would be deserts instead of the reasonably good pine stands we see today. Pines, heavily inoculated with mycorrhizae in the nursery before planting, are able to grow even on the land devastated by smelter fumes at Copperhill (see Chapter 3). (For more on the practical importance of mycorrhizal associations, see Ruehle and Marx 1979.)

Lichens

One whole group of plants, the **lichens**, is composed of mutualistic algae and fungi that are so closely associated that botanists find it convenient to consider the association a single species. It is probable that this mutualism evolved not only from commensalism and cooperation but also from parasitism. In some primitive lichens, for example, the fungi actually penetrate the algal cells, as shown in Figure 10A, and are thus essentially parasites of the algae. In the more advanced species the fungal mycelia do not break into the algal cells, but the two live in close harmony (Figures 10B and 10C). The fungus absorbs the photosynthate that leaks out of the

(A)

(B)

FIGURE 9
Two types of mycorrhizae (fungus–root mutualisms). (A) Young pine
seedlings devoid of mycorrhizae (left) and with a prolific development of
ectomycorrhizae (right). (B) Vesicular–arbuscular (VA, or endo-) mycorrhizae, showing fungal mycelia and vesicles inside root cells. (Photographs courtesy of S. A. Wilde.)

algal cells, and in return, the alga receives the support and protection of
the fungus. So successful is the partnership that lichens are able to live in
the harshest physical environments, such as granite outcrops and Arctic
tundras.

FIGURE 10
An evolutionary trend from parasitism to mutualism in lichens. In some primitive lichens, the fungal hyphae actually penetrate the algal cells (A) in an essentially parasitic relationship. In more advanced lichen species, the two organisms live in greater harmony for mutual benefit, as in (B), where the fungal hyphae intermingle with the algal filaments, and (C), where the fungal hyphae are pressed close against the algal cells but do not penetrate them.

It is evident that mutualism has special survival value when resources become tied up in the biomass, as in a mature forest, or when soil or water is nutrient-poor, as in some coral reefs or rain forests. Like corals and other highly organized heterotroph–autotroph mutualistic complexes, lichens are well adapted to natural scarcities and stress, but are very vulnerable to pollution stress, especially air pollution. In the effort to restore the landscape at Sudbury, Ontario, that was devastated by air pollution (mentioned in Chapter 3), the return of lichens is a welcome sign that the restoration is working.

The mutualistic lichen lifestyle has had at least five independent origins in different branches of the fungus family tree; at least 20 percent of all fungal species are lichens (Gargas et al. 1995). Such multiple origins demonstrate that mutualism may be just as important as competition in evolution, as suggested by Kropotkin almost a century ago. We will consider the theoretical aspects of the evolution of cooperation and mutualism, and their relevance to human affairs, in the next chapter.

Network Mutualism

The indirect effects of species on one another may be as important as their direct interactions, and may contribute to **network mutualism**. When food chains function in food web networks, the organisms at each end of a trophic series—for example, plankton and bass in a pond—do not interact directly, but they benefit each other indirectly. Bass benefit by eating planktivorous fish, which are supported by the plankton, while plankton benefit when bass reduce the population of the plankton's predators. Accordingly, there are both negative (predator–prey) and positive (mutualistic) interactions in a food web network (Wilson 1986; Patten 1991).

Because of "reward feedbacks," as discussed in Chapter 4, and because of the tendency for the severity of negative interactions to decrease with time, as discussed earlier in this chapter, it is not too far-fetched to consider whole food chains as mutualistic (Odum and Biever 1984). In a recent study of an algae–herbivore relationship, Sterner (1986) found that the algae grew better when grazed because of the nitrogen regenerated by the grazer. (For general reviews of mutualism and its importance, see Boucher 1985; Boucher et al. 1982; and Keddy 1990.)

The ultimate reality is that all positive and negative two-species interactions operate together in food webs at the community and ecosystem levels. The energetics of food chains, as detailed in Chapter 3, combined with what have come to be known as "top-down" and "bottom-up" processes, make the food web a functional system that is more than just a collection of species interactions. Top-down control, which includes reward feedback, refers to the role of "upstream" components—for example, herbivore control over plants and predator control over herbivores. Bottom-up control refers to the role of nutrients and other physical factors in determining primary production. Ecologists argue over which is most important in a given situation, but most now agree that both are involved to varying degrees in any and all natural situations (Hunter and Price 1992; Polis 1994; de Ruiter et al. 1995; Krebs et al. 1995).

Suggested Readings

*Abernathy, V. 1993. The demographic transition revisited: Lessons from foreign and U.S. immigration policy. *Ecol. Econ.* 8: 255–257.

*Acharya, A. 1995. Small islands awash in a sea of troubles. *Worldwatch* 6(6): 24–33.

*Alexander, M. 1981. Why microbial parasites and predators do not eliminate their prey and hosts. *Annu. Rev. Microbiol.* 35: 113–133.

*Allee, W. C. 1951. *Cooperation among Animals, with Human Implications.* Henry Schuman, New York.

Allee, W. C. 1958. *The Social Life of Animals.* Beacon Press, Boston.

*Allen, M. F. 1991. *The Ecology of Mycorrhizae.* Cambridge University Press, New York.

*Anagnostakis, S. L. 1982. Biological control of chestnut blight. *Science* 215: 466–471.

*Anderson, R. M., and R. M. May. 1982. Coevolution of hosts and parasites. *Parasitology* 85: 411–426.

Ayala, F. J. 1972. Competition between species. *Am. Sci.* 60: 348–357.

*Biswas, A. K. 1993. Land resources for sustainable agricultural development in Egypt. *Ambio* 22: 556–580.

*Bongaarts, J. 1994. Population policy options in the developing world. *Science* 263: 771–776.

*Boucher, D. H., ed. 1985. *The Biology of Mutualism*. Oxford University Press, New York.

*Boucher, D. H., S. James, and K. H. Keeler. 1982. The ecology of mutualism. *Annu. Rev. Ecol. Syst.* 13: 315–347.

*Brown, L. R., and J. L. Jacobson. 1986. *Our Demographically Divided World*. Worldwatch Paper no. 74. Worldwatch Institute, Washington, D.C.

Burkholder, P. R. 1952. Cooperation and conflict among primitive organisms. *Am. Sci.* 40: 601–631. (Considers the nine possible types of interaction as first suggested by E. F. Haskell in *Main Currents in Modern Thought* 7: 45–51, 1949.)

Calhoun, J. B. 1962. Population density and social pathology. *Sci. Am.* 206(2): 139–148.

*Carpenter, J. R. 1940. Insect outbreaks in Europe. *J. Anim. Ecol.* 9: 108–147.

*Catton, W. R. 1987. The world's most polymorphic species: Carrying capacity transgressed two ways. *BioScience* 37: 413–419.

*Catton, W. R. 1995. Backing into the future. *Focus: Carrying Capacity Network* 5(1): 41–46. Washington, D.C.

*Christensen, A. M., and J. McDermott. 1958. Life history and biology of the oyster crab, *Pinnotheres ostreum*. *Biol. Bull.* 144: 146–179.

*Coley, P. D., J. P. Bryant, and F. S. Chapin III. 1985. Resource availability and plant antiherbivore defense. *Science* 230: 895–899.

Colinvaux, P. A. 1982. *Why Big Fierce Animals are Rare: An Ecologist's Perspective*. Princeton University Press, Princeton, NJ. (A delightfully written and provocative book.)

*Connell, J. H. 1961. The influence of interspecific competition and other factors on the distribution of the barnacle *Chthamalus stellatus*. *Ecology* 42: 710–723.

Dawkins, R. 1986. *The Blind Watchmaker*. Norton, New York. (Good review of evolutionary theories.)

Deevey, E. S. 1960. The human population. *Sci. Am.* 203(3): 195–204.

*den Boer, P. J. 1986. The present status of the competition exclusion principle. *Trends Ecol. Evol.* 1: 25–28.

*de Ruiter, P. C., A. M. Neutel, and J. C. Moore. 1995. Energetics, patterns of interactive strengths and stability in real ecosystems. *Science* 269: 1257–1260.

*Dindal, D. L. 1975. Symbiosis: Nomenclature and proposed classification. *Biologist* 54(4): 129–142.

Ehrlich, P. R. 1968. *The Population Bomb*. Ballantine Books, New York.

*Ehrlich, P. R., and A. E. Ehrlich. 1992. The most overpopulated nation. *Clearinghouse Bulletin, Carrying Capacity Network* 2(8): 1–3, 7.

Ehrlich, P. R., and H. A. Mooney. 1983. Extinction, substitution, and ecosystem services. *BioScience* 33: 248–254. (Extinction of key wild species may result in loss of vital life-support services to humans.)

Enke, S. 1969. Birth control for economic development. *Science* 164: 798–802. (Reducing human fertility can raise per capita income in underdeveloped countries.)

*Errington, P. L. 1946a. Predation and vertebrate populations. *Q. Rev. Biol.* 21: 144–177, 221–245.

*Errington, P. L. 1946b. Vulnerability of bob-white population to predation. *Ecology* 15: 112–127.

*Ewald, P. W. 1983. Host–parasite relations, vectors, and the evolution of disease severity. *Annu. Rev. Ecol. Syst.* 14: 465–485.

*Freedman, D., and B. Berelson. 1974. The human population. *Sci. Am.* 231(3): 30–39.

Galle, O. R., W. R. Gove, and J. M. McPherson. 1972. Population density and pathology: What are the relations for man? *Science* 176: 23–30. (Evidence from one city suggests high density may be linked with pathological behavior, as found in John B. Calhoun's animal experiments; see Calhoun 1962.)

*Gargas, A., P. T. DePriest, M. Grube, and A. Tehler. 1995. Multiple origins of lichen symbiosis in fungi suggested by SSUrDNA phylogeny. *Science* 268: 1492–1495.

*Gause, G. F. 1932. Ecology of populations. *Q. Rev. Biol.* 7: 27–46.

*Gordon, D. M. 1995. The development and organization of an ant colony. *Am. Sci.* 83: 50–57.

*Gould, S. J. 1988. Kropotkin was no crackpot. *Nat. Hist.* 97(7): 12–21.

*Hanski, I. 1989. Metapopulation dynamics: Does it help to have more of the same? *Trends Ecol. Evol.* 4: 113–114.

*Hardin, G. 1960. The competitive exclusion principle. *Science* 131: 1292–1297.

Hardin, G. 1986. Cultural carrying capacity. *BioScience* 36: 599–606.

*Harper, J. L., and J. N. Clatworthy. 1963. The comparative biology of closely related species of clover in mixed and pure culture. *J. Exp. Bot.* 14: 172–190.

*Harris, L. D. 1984. *The Fragmented Forest: Island Biogeography Theory and Preservation of Biotic Diversity*. University of Chicago Press, Chicago.

*Hunter, M. D., and P. W. Price. 1992. Playing chutes and ladders: Heterogeneity and the relative roles of bottom-up and top-down forces in natural selection. *Ecology* 73: 724–732.

Hutchinson, G. E. 1978. *An Introduction to Population Ecology*. Yale University Press, New Haven, CT.

Johnson, D. D., and R. Johnson. 1989. *Cooperation and Competition: Theory and Research*. Interaction Book Co., Edina, MN. (Social scientists find that cooperative learning strategies—when students work together in small groups and help and teach each other—are far more effective in developing student talent than traditional competitive approaches.)

*Keddy, P. 1990. Is mutualism really irrelevant to ecology? *Bull. Ecol. Soc. Am.* 71(2): 101–102.

*Klein, D. R. 1968. The introduction, increase, and crash of reindeer on St. Matthew Island. *J. Wildl. Mgmt.* 32: 350–367.

*Krebs, C. J., S. Boutin, R. Boonstra, R. E. Sinclair, J. N. Smith, M. R. T. Dale, K. Martin, and R. Turkington. 1995. Impact of food and predation on snowshoe hare cycles. *Science* 269: 1112–1115. (Enclosure experiments suggest that bottom-up, i.e., food, and top-down, i.e., predation, processes interact to generate cycles.)

*Kropotkin, P. 1902. *Mutual Aid: A Factor of Evolution*. William Heinemann, London. (Reprinted in 1935, Extending Horizons Books, Boston.)

*Levin, S., and D. Pimentel. 1981. Selection of intermediate rates of increase in parasite–host systems. *Am. Nat.* 117: 308–315.

*Levins, R. 1969. Some demographic and genetic consequences of environmental heterogeneity for biological control. *Bull. Entomol. Soc. Am.* 15: 237–240.

*Malthus, T. R. 1798. *An Essay on the Principle of Population*. Johnson, London. (Reprinted in Everyman's Library, 1914.)

Mauldin, W. P. 1980. Population trends and prospects. *Science* 209: 148–157.

*McCullough, D. R. 1979. *The George Reserve Deer Herd: Population Ecology of a K-Selected Species*. University of Michigan Press, Ann Arbor.

*McNamara, R. 1982. Demographic transition theory. In *International Encyclopedia of Population*, vol. 1. Prentice-Hall, Englewood Cliffs, NJ.

*Meyers, J. H. 1993. Population outbreaks in forest Lepidoptera. *Am. Sci.* 81: 240-251.

Myers, N. 1979. *The Sinking Ark: A New Look at the Problem of Disappearing Species*. Pergamon Press, Elmsford, NY.

*Myers, N. 1983. *A Wealth of Wild Species: Storehouse for Human Welfare*. Westview Press, Boulder, CO.

*National Academy of Sciences. 1971. *Rapid Population Growth: Consequences and Policy Implications*. Johns Hopkins University Press, Baltimore. (Concludes that rapid human population growth has more economic disadvantages than advantages because costly problems develop faster than solutions.)

*National Research Council. 1986. *Population Growth and Economic Development: Policy Questions*. National Academy Press, Washington, D.C. (Rapid population growth, while not the cause of all the problems in the Third World, is more likely to impede progress than to promote it.)

*Newell, S. J., and E. J. Tramer. 1978. Reproductive strategies in herbaceous plant communities during succession. *Ecology* 59: 228–234.

*Norton, B. G. 1986. *The Preservation of Species: The Value of Biological Diversity*. Princeton University Press, Princeton, NJ.

*Odum, E. P. 1983. Population ecology. Chapters 6 and 7 in *Basic Ecology*. Saunders College Publishing, Philadelphia.

*Odum, E. P., and L. J. Biever. 1984. Resource quality, mutualism, and energy partitioning in food chains. *Am. Nat.* 124: 360–376.

Odum, H. T. 1976. Energy quality and the carrying capacity of the earth. *Trop. Ecol.* 16: 1–8. (Human carrying capacity depends on the quality as well as the quantity of available energy.)

Owens, D. F., and R. G. Wiegert. 1976. Do consumers maximize plant fitness? *Oikos* 27: 489–492.

*Paine, R. T. 1966. Food web diversity and species diversity. *Am. Nat.* 100: 65–75.

*Park, T. 1954. Experimental studies on interspecific competition. *Physiol. Zool.* 27: 177–238.

Park, T. 1962. Beetles, competition, and populations. *Science* 138: 1369–1375.

*Patten, B. C. 1991. Network ecology: Indirect determination of the life-environment relationship in ecosystems. In *Theoretical Studies in Ecosystems: The Network Perspective*, eds. M. Higashi and T. P. Burns, pp. 288–351. Cambridge University Press, Cambridge.

*Peakall, O. B., and P. N. Whit. 1976. The energy budget of an orb web-building spider. *Biochem. Physiol.* 54: 187–190.

Perry, N. 1983. *Symbiosis: Close Encounters of the Natural Kind*. Sterling, New York.

*Pimentel, D. 1968. Population regulation and genetic feedback. *Science* 159: 1432–1437. (Evolutionary tendency for severe negative interactions to be reduced or to become positive.)

*Pimentel, D., and F. A. Stone. 1968. Evolution and population ecology of parasite–host systems. *Canad. Entomol.* 100: 655–662.

*Platt, A. E. 1996. Infecting ourselves: How environmental and social disruptions trigger disease. Worldwatch Paper 129, Worldwatch Institute, Washington, D.C.

*Polis, G. A. 1994. Food webs, trophic cascades, and community structure. *Aust. J. Ecol.* 19: 121–136.

Pulliam, H. R., and M. M. Haddad. 1994. Human population growth and the carrying capacity concept. *Bull. Ecol. Soc. Am.* 75(3): 141–157.

Quinn, J. A. 1978. Plant ecotypes: Ecological or evolutionary units. *Bull. Torrey Bot. Club* 105: 58–64.

*Riechert, S. E. 1981. The consequences of being territorial: Spiders, a case study. *Am. Nat.* 117: 871–892.

*Ruehle, J. L., and D. H. Marx. 1979. Fiber, food, fuel, and fungal symbionts. *Science* 206: 419–422. (Importance of mycorrhizae in food, fiber, and fuel production.)

Scientific American. 1974. Special issue on the human population. 231(3).

Seaman, G. A. 1952. The mongoose and Caribbean wildlife. *Trans. N. Am. Wildl. Conf.* 17: 188–197.

Selye, H. 1973. The evolution of the stress concept. *Am. Sci.* 61: 692–699. (Stress as a nonspecific response of the body to demands made on it is an important medical concept in relation to population pressure and toxic substances in the environment.)

Soulé, M. E., ed. 1986. *Conservation Biology: The Science of Scarcity and Diversity.* Sinauer Associates, Sunderland, MA.

*Sterner, R. W. 1986. Herbivores' direct and indirect effects on algal populations. *Science* 231: 605–607.

*Stoddard, H. L. 1936. Relation of burning to timber and wildlife. *Proc. 1st N. Am. Wildl. Conf.* 1: 1–4.

*Teitelbaum, M. S. 1975. Relevance of demographic transition theory for developing countries. *Science* 188: 420–425.

*Todes, D. P. 1989. Kropotkin's theory of mutual aid. Chapter 7 in *Darwin without Malthus: The Struggle for Existence in Russian Evolutionary Thought.* Oxford University Press, New York.

Toft, C. A., A. Aeschlimann, and L. Bolis. 1991. *Parasite–Host Associations: Coexistence or Conflict?* Oxford University Press, New York.

*Werner, E. E., and D. J. Hall. 1974. Optimal foraging and the size selection of prey by the bluegill sunfish (*Lepomis macrochirus*). *Ecology* 55: 1042–1052.

*Wilde, S. A. 1968. Mycorrhizae and tree nutrition. *BioScience* 18: 482–484.

*Wilson, D. S. 1986. Adaptive indirect effects. Chapter 26 in *Community Ecology,* eds. J. Diamond and T. J. Case, pp. 437–444. Harper & Row, New York.

*Wilson, E. O., ed. 1988. *Biodiversity.* National Academy Press, Washington, D.C.

Wilson, E. O. 1991. Ants. *Bull. Am. Acad. Arts Sci.* 45(3): 13–23. (Ants are the dominant "little-sized organisms," with greater biomass than any other insect group. Ant colonies practice "programmed demography" in that reproduction is reduced when the colony becomes large.)

*Indicates references cited in this chapter.

7

Development and Evolution

Biotic communities go through a youth-to-maturity development process analogous to the growth and development of an individual organism, but the patterns and controls are quite different, as we will see. Community development over the short term (1000 years or less) is widely known as **ecological succession**, but is perhaps better thought of as **ecosystem development** because it is an active process involving changes in both the organisms and the physical environment. Changes over geological time (millions of years) fall under the heading of **organic evolution**.

You are probably already aware of ecological succession, since it goes on continually all around you in the landscape. But you may not have been aware that there are definite patterns in these changes which, in the absence of major disturbances, are predictable. When an area becomes available for community development (as, for example, when a crop field is abandoned and left for nature to redevelop), opportunistic plants and animals colonize it in a series of temporary, or pioneer, communities called **seral stages**. Gradually, more permanent communities develop until a mature or **climax stage** that is metabolically balanced (i.e., in which primary production, P, roughly approximates maintenance or respiration, R) takes over. Its biotic composition is determined by the regional climate and the local substratum, topography, and water conditions.

An Example: Old Field Succession

The pattern of ecological succession on abandoned cropland in the Piedmont region of Georgia is illustrated in Figure 1; it shows the changes in vegetation that are a part of the successive biotic communities that develop over time in the absence of major natural or human disturbances. Annual "weedy" plants, such as crabgrass and ragweed, are the first to colonize the abandoned plowed ground, followed in a few years by perennial forbs and grasses (aster, goldenrod, and broomsedge grass, for example), then shrubs and pine seedlings. A closed-canopy pine forest develops and persists for 100 years or so, to be gradually replaced by a shade-tolerant hardwood forest (oaks and hickories are dominants in the Piedmont climax forest). As shown in Figure 1, a bird succession accompanies the vegetative changes, with open country and forest-edge species being replaced by forest-interior species as the forest matures. A pattern similar to the one in Figure 1 can be expected in any area where some kind of forest vegetation is the climax stage, but the species of plants and animals that take part in the development will vary with the topography, climate, and geographic region (recall the discussion of "ecological equivalents" in Chapter 3).

Successional Theory: A Brief History

Although ecological succession was first described by Europeans (especially Eugenius Warming in 1895), it was Frederic E. Clements, born and bred on the Nebraska prairies at the turn of the twentieth century, who pioneered the field. Clements viewed the landscape as a dynamic entity with a life history of its own. He and his wife Edith, a skilled botanist, were tireless fieldworkers who traveled far and wide investigating the history, structure, and composition of vegetation. In his monograph *Plant Succession: An Analysis of the Development of Vegetation*, published in 1916, Clements pictured the biotic community as a "superorganism" with a

FIGURE 1

The general pattern of ecological succession on abandoned farmland in the southeastern United States. The diagram shows four stages in the life form of the vegetation (grassland, shrubs, pines, and hardwoods); the bar graph shows changes in the songbird population that accompany the changes in autotrophs. A similar pattern will be found in any area where the climax stage is forest, but the species of plants and animals that take part in the development series will vary according to the climate or topography of the area. (After Johnston and Odum 1956.)

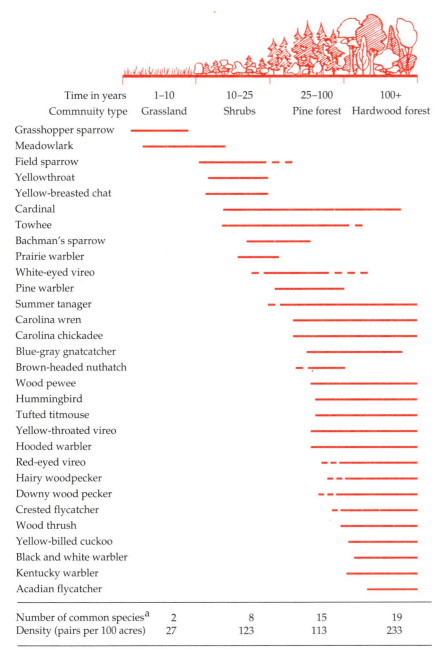

Time in years	1–10	10–25	25–100	100+
Commnuity type	Grassland	Shrubs	Pine forest	Hardwood forest

Grasshopper sparrow

Meadowlark

Field sparrow

Yellowthroat

Yellow-breasted chat

Cardinal

Towhee

Bachman's sparrow

Prairie warbler

White-eyed vireo

Pine warbler

Summer tanager

Carolina wren

Carolina chickadee

Blue-gray gnatcatcher

Brown-headed nuthatch

Wood pewee

Hummingbird

Tufted titmouse

Yellow-throated vireo

Hooded warbler

Red-eyed vireo

Hairy woodpecker

Downy wood pecker

Crested flycatcher

Wood thrush

Yellow-billed cuckoo

Black and white warbler

Kentucky warbler

Acadian flycatcher

Number of common species[a]	2	8	15	19
Density (pairs per 100 acres)	27	123	113	233

[a]A common species is arbitrarily designated as one with a density of 5 pairs per 100 acres or greater in one or more of the four community types.

development parallel to that of an individual organism. Furthermore, he believed that for a given region, there was only one climax stage toward which all vegetation was developing, however slowly (this was subsequently known as the "monoclimax" theory, in contrast to a "polyclimax" theory that allows for many possible terminal stages).

Quite a different view of the plant community was presented by Clements's contemporary, Herbert A. Gleason, in a paper entitled "The Individualistic Concept of the Plant Association," published in 1926. Gleason was skeptical that there is any organizational strategy at the community level; rather, he argued, ecological succession results from the interaction of individuals and species as they struggle to occupy and hold space. Both Clements and Gleason assumed that plants were entirely or mostly responsible for successional changes. We now know that animals and microorganisms play vital roles, as was first shown by pioneer animal ecologist Victor E. Shelford. In 1939, Clements and Shelford coauthored a book entitled *Bio-ecology* that attempted, rather unsuccessfully, to describe plant–animal interactions at the community level. Not until ecosystem energetics began to be studied could the role of autotroph––heterotroph interactions be understood.

The eminent Spanish ecologist Ramón Margalef (1968) was among the first to demonstrate that ecosystem development involves a fundamental shift in energy allocation between production (P) and respiration (R), with P greater than (or sometimes less than) R in the early stages and P equal to R in the climax stage. Such a shift is definitely an ecosystem-level strategy.

As is so often the case with theories and controversies in science, neither extreme proved to be acceptable as a sole explanation for phenomena. Both holistic and individualistic processes appear to be involved in community development. Communities and ecosystems are not "superorganisms," but they are nonequilibrium systems that have the capacity for self-organization, as was detailed in Chapter 4. Current theory holds that ecosystem development results from modification of the physical environment by the biotic community acting as a whole (the holistic component); the interaction of competition and coexistence between component populations (the individualistic component); and a shift in energy flow from production to respiration as more and more of the available energy is required to support the increasing organic structure (the community metabolism component).

Generally accepted today is the concept that ecological succession is a two-phased process. Early or **pioneer** stages tend to be random (i.e., stochastic) as opportunistic species colonize, but later stages tend to be more self-organized (i.e, deterministic).

Types of Succession

Succession that begins on a sterile site where conditions for life are not at first favorable (for example, a newly exposed sand dune or lava flow) is called **primary succession**. As might be expected, development can be very slow on such sites. Olson (1958), who reexamined the plant succession first described by H. C. Cowles in 1899 on the sand dunes left behind when Lake Michigan retreated northward, has estimated that about 1000 years are required for nature to develop a climax hardwood forest starting with bare dunes, assuming no wind, water overwash, bulldozers, or other disturbance. Of course, the probability is high that some kind of disturbance will interfere with the development process during such a long time period. Fortunately, a portion of these dunes is included in the Indiana Dunes National Park, preserving this field laboratory for the continued study of primary succession.

In contrast, the term **secondary succession** is used for community development on sites previously occupied by well-developed communities or where nutrients and other conditions are favorable, such as abandoned cropland, plowed grasslands, cut forests, or new ponds. Secondary succession can be rapid, with mature stages developing in a few decades in a grassland or aquatic environment, or in less than 500 years in a forest (see Figure 1). Soil development, which of course is a major part of terrestrial community development, was discussed within this time frame in Chapter 5.

It is important to distinguish between **autotrophic succession** and **heterotrophic succession**. The former is the most common type in nature. It begins in a predominantly inorganic environment and is characterized by early and continued dominance of green plants (autotrophs). In contrast, heterotrophic succession is characterized by early dominance of heterotrophs; it occurs in special cases where the environment is primarily organic, as, for example, in a stream heavily polluted with sewage or, on a smaller scale, in a rotting log. Energy is maximum at the beginning and declines as succession occurs, unless additional organic matter is imported or until an autotrophic regime takes over. In contrast, energy flow does not necessarily decline in autotrophic succession, but is usually sustained or increased.

Models of Ecosystem Development

Figure 2 is a general systems model of succession, based on the concept that internal or **autogenic inputs**, operating more or less continuously,

FIGURE 2
A systems model of ecological succession.

and periodic external or **allogenic inputs** both affect the progress of a system developing toward climax. In theory, the autogenic forces tend to drive the system toward an equilibrium, with a balance between P and R and a stabilized species composition. By contrast, strong allogenic inputs tend to disrupt progress toward equilibrium and set back the succession to a younger stage, as happens when a forest is storm-damaged or cut over, or when sewage is dumped into a pond. Sometimes an allogenic input accelerates rather than hinders development toward equilibrium.

Figure 3, an energy flow model, shows the basic change in energy partitioning between P and R mentioned above. Early in the development of

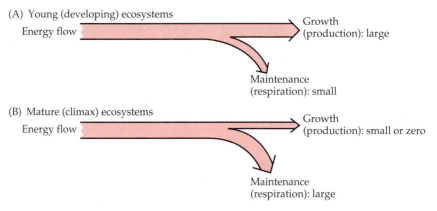

FIGURE 3
An energy flow model of ecological succession, contrasting energy partitioning in (A) developing systems and (B) mature systems.

an ecosystem, when there is not much biomass to support, a large part of the available energy goes into new growth (production). But as the organic structure builds up, more and more energy is required to maintain this structure and dissipate disorder, and so less is available for production, as discussed in Chapter 4. This shift in energy use has parallels in the development of human societies, and it greatly affects our attitudes about how our environment should be treated, as we shall see.

Microcosm Models

Ecosystem development on a microcosmic scale can be set in motion by placing glass flasks under a good light source, as in a laboratory growth chamber. The flasks are half-filled with a culture medium containing a balance of the inorganic salts necessary for life, then inoculated with samples of water and sediments from a pond. It is good practice to cross-inoculate the flasks to make sure a variety of small organisms or their propagules (both plant and animal) get into each container.

Figure 4A shows how three properties, namely, production by photosynthesis (P), respiration (R), and biomass (B), change with time in a microcosm model. The basic pattern of succession in the microcosm parallels that which occurs in the longer, open-system development of a forest, shown in Figure 4B. During the first few weeks in the microcosm, the autotrophs (algal cells) introduced with the pond water exploit the temporarily unlimited nutrients and undergo rapid growth. Small heterotrophs (bacteria, protozoa, nematodes, crustaceans, and so on) respond likewise, so that the total volume of living material, or biomass, increases rapidly. During this youthful growth stage, the total, or gross, production (P_g) exceeds the total respiration (R) so that $P/R > 1$, leaving substantial net production (P_n) to accumulate as biomass.

As resources, such as space and nutrients, reach saturation use in the closed system, the rate of production becomes limited by the rate of decomposition and regeneration of nutrients. A pulsing climax state then develops in which production on average balances respiration, so that $P = R$ or $P/R = 1$. At this stage there is little or no net production, and no further increase in biomass. In appearance, the culture goes from bright green to yellow-green in color, with detritus and detritivores becoming more prominent in the mature stage. Climax cultures can continue indefinitely, but if the diversity in the original inoculation was low, the culture may age and die due to lack of organisms able to carry out key processes. New successions can be set in motion at any time by inoculations from old cultures into new media, or by adding new media to old cultures.

As indicated above, succession may proceed either from the extreme autotrophic condition, as in the cultures just described, or from the

(A) Microcosm succession

(B) Forest succession

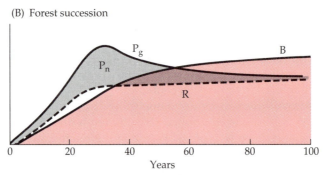

FIGURE 4
(A) A microcosm model of ecological succession, compared with (B) the energetics of forest succession. P_g, gross primary production; P_n, net primary production; R, respiration; B, biomass.

extreme heterotrophic condition, in which R exceeds P (or where P may be zero). An interesting microcosm model of heterotrophic succession is the hay infusion culture often used for growing protozoans and other microscopic animals. If a quantity of dried hay is boiled, and the solution allowed to stand a few days in the dark, a thriving culture of heterotrophic bacteria develops. If pond water containing seed stocks of small animals is added, a succession of species can be observed for about a month. Usually small flagellates called monads appear first, followed by ciliated protozoa such as *Paramecium* and *Colpoda*; changes then come more slowly, with specialized ciliates (such as *Hypotricha* and *Vorticella*), amoebas, and rotifers reaching successive peaks of abundance. If no algae are present, or if the culture is kept in the dark and no fresh hay infusion is added, the microcosm will run down, since all the organisms will die for lack of food, the original organic matter having been used up.

Thus, the two types of succession can be contrasted on a small scale in the laboratory, or as a class exercise in a course in ecology. The cultures demonstrate what happens in the early stages of autotrophic succession in a new pond or artificial lake, and in the heterotrophic succession that occurs after sewage is discharged into a pond or stream. (For more on microcosms as models for succession, see Cooke 1967; Gorden et al. 1969; and Beyers and H. T. Odum 1993.)

In general, laboratory microcosms are too small and too closed and do not contain enough diversity (physical and biological) to reveal all the important features of ecosystem development. So we need to develop a more comprehensive model.

A Tabular Model for Autogenic Ecosystem Development

A summary of the important changes in community structure and function that are to be expected during autogenic succession, as revealed by the study of the large, open systems of nature, is shown in Table 1. Trends in ecological development are grouped under several headings. Although ecologists have studied succession in many parts of the world, most of the emphasis to date has been on descriptive aspects such as changes in species composition or nutrient levels; only recently have functional aspects also been considered. Consequently, some of the items listed in Table 1, especially those at the bottom of the table, must be considered provisional or hypothetical, i.e., not fully verified by adequate field data or experimental testing. Five aspects of the model seem most significant and require a bit more explanation.

First, the species of plants and animals (and microorganisms) present change with successive stages in ecosystem development, a process that has been called **relay floristics and faunistics**. When the occurrence (or better still, density) of species is plotted against time, a characteristic stairstep graph is obtained, as illustrated in Figure 1. Such a pattern is usually apparent whether we are considering a specific taxonomic group, such as birds, or a trophic group, such as producers or herbivores. Typically, some species in the series have wider habitat tolerances than others and persist over longer periods of time, as, for example, pines and cardinals in the Piedmont. In general, the more species in a group (whether taxonomic or ecological) that are geographically available for colonization, the more restricted will be the occurrence of each species in the time sequence, due to the competition–coexistence interactions discussed in the preceding chapter.

Second, diversity tends to increase with succession, especially in primary succession and in the early stages of secondary succession. Quite

Table 1   A Tabular Model for Ecological Succession of the Autogenic, Autotrophic Type

Ecosystem characteristic	Trend in ecological development Early stage → Climax Youth → Maturity Growth stage → Steady state
Community Structure	
Species composition	Changes rapidly at first, then more gradually (relay floristics and faunistics)
Size of individuals	Tends to increase
Species diversity	Increases initially, then stabilizes or declines in older stages as size of individuals increases
Total biomass (B)	Increases
Nonliving organic matter	Increases
Energy Flow (Community Metabolism)	
Gross production (P)	Increases during early phase of primary succession; little or no increase during secondary succession
Net community production (yield)	Decreases
Community respiration (R)	Increases
P/R ratio	P > R to P = R
P/B ratio	Decreases
B/P and B/R ratios (biomass supported/unit energy	Increases
Food chains	From linear chains to more complex food webs
Biogeochemical Cycles	
Mineral cycles	Become more closed
Turnover time and storage of essential elements	Increases
Internal cycling	Increases
Nutrient conservation	Increases
Natural Selection and Regulation	
Growth form	From r-selection (rapid growth) to K-selection (feedback control)
Life cycles	Increasing specialization, length, and complexity
Symbiosis (living together)	Increasingly mutualistic
Entropy	Decreases
Information	Increases
Overall efficiency of energy and nutrient utilization	Increases

often, the highest diversity occurs in the intermediate stages of succession (Sousa 1984). However, trends may differ with different taxonomic and trophic groups. For example, the diversity of autotrophs may reach a maximum early in forest succession, then decline as trees get larger, while the diversity of heterotrophs may continue to increase into the climax stage. The interplay of opposite trends makes it difficult to generalize in regard to diversity. Increases in the size of individual organisms and increases in competition tend to reduce diversity, while increases in organic structure and variety of habitats tend to increase it. Periodic disturbances (e.g., fire, storms, and predators) often increase species diversity by opening up areas for colonization by species that are unable to survive in the undisturbed community. The "intermediate disturbance hypothesis" was discussed briefly in Chapter 3. If disturbances continue, the succession is set back to earlier stages or maintained in a nonequilibrium state (a related concept, pulse-stabilized subclimax, will be discussed later in this chapter.)

Third, biomass and the standing crop of organic matter increase with succession. In both aquatic and terrestrial environments, the total amount of living matter (biomass) and decomposing organic material (detritus and humus) increases with time until equilibrium is approximated (input = output). Dissolved organic matter (DOM), which leaks out of decomposing matter as well as from living cells, accumulates in increasing amounts and varieties. These "extrametabolites" not only power microbial food chains, but some products also act as inhibitors (antibiotics) and growth promoters (vitamins, for example) that affect growth and species composition, as was briefly discussed in Chapter 4. Creating an increasingly organic environment is one of the main ways the community facilitates a succession of species.

Fourth, a decrease in net production and a corresponding increase in respiration are two of the most striking and important trends in succession. These changes, which were explained earlier in this chapter, are shown graphically in Figure 3.

Fifth, lifestyles and life histories change during succession. Not only do species come and go during succession, but adapted lifestyles tend to shift from *r*-selected to *K*-selected (recall the discussion of *r*- and *K*-selection in Chapter 6). Species found in pioneer or early seral stages are often *r*-strategists in that they exhibit high reproductive rates and have simple life histories. In contrast, the capacity to live in a crowded world of limited resources, a *K*-strategy, has greater survival value in the climax stage. Larger body sizes or increased storage capacity, more specialized niches, longer and more complex life histories, and more cooperation between species (mutualism) are attributes that become more important than reproductive capacity as the ecosystem matures. If a single species is to survive in communities that vary all the way from pioneer systems to mature systems (and very few are able to do so), then dramatic changes must be made in its lifestyle. Humans face the problem of readjusting our lifestyles as we move from uncrowded pioneer societies to mature, crowded societies.

We now come to the hypothetical, controversial, and hard-to-test part of the story of ecological succession. While most of the trends outlined in Table 1 are well documented and accepted by ecologists, the "how" and the "why" remain as debatable as in Clements's and Gleason's time. Succession is not "goal-oriented" development with a centralized set-point control (as is the development of an individual organism), but is the result of the inherent capacity of nonequilibrium systems to self-organize as a result of a diffuse network of subsystem feedbacks (see Figure 11 in Chapter 3). Brooks and Wiley (1986) contend that the outcome of the second law of thermodynamics is self-organization, with the result that living systems exhibit increasing complexity because of, not in spite of, the expense of entropy. The overall "strategy" involves decreasing entropy (disorder), increasing information (order), increasing the ecosystem's ability to survive perturbations (resistance stability), and increasing efficiency of energy and nutrient utilization. Many ecologists do not accept this hypothesis. Much depends on the outcome of the current debate on the mechanisms of evolutionary change, which we shall touch on briefly later in this chapter.

Based on network theory, Robert Ulanowicz (1980) suggests that the natural tendency of ecosystem development is in the direction of **ascendancy** (i.e., the system "ascends" to a more mature configuration). He believes that the development process can, in theory at least, be quantified by encapsulating major trends (increased species richness, greater cycling and feedback, greater niche specialization, increased metabolism; see Table 1) in an "index of network ascendancy" that is a product of systems activity and trophic (food web) organization.

The Time Factor and Allogenic Forces

While the changes shown in Figure 4 and Table 1 seem independent of geographic location or type of ecosystem, the physical environment and allogenic forces strongly affect the time required for succession—that is, whether the time scale (the x-axis in Figures 1 and 4) is measured in weeks, months, or years—and the relative stability or persistence of the climax stage. In open-water systems, as in cultures, the community is able to modify its physical environment to only a small extent; consequently, succession, if it occurs at all, is brief, perhaps lasting for only a few weeks. A climax, if it can be said to occur, has a limited life span. Margalef (1968) summarizes his observations of the changes that occur in a seasonal succession gradient in the coastal water column as follows:

1. The average size of cells and the relative abundance of mobile forms among the phytoplankton increase.
2. The rate of productivity slows down.
3. The chemical composition of the phytoplankton changes, as exemplified by the change in plant pigments from bright green to yellow-green.
4. The composition of the zooplankton shifts from passive filter feeders to more active and selective hunters, in response to a shift from numerous small food particles to scarcer food concentrated in larger units in a more organized (stratified) environment.
5. In the later stages of succession, total energy transfer may be lower, but its efficiency seems to be improved.

Note that these observations parallel fairly closely the trends shown in Table 1.

In a forest ecosystem, to take the other extreme, the community is able to modify the physical environment extensively. A large biomass accumulates, and community structure continues to change in a predictable manner over a long period of time—unless or until the autogenic processes are interrupted by severe disturbances such as storms. To predict or model forest succession, disturbance regimes have to be included in the space–time domain that is being considered (Shugart 1984). In a study of the vegetative history of a small area of the Harvard Forest in Massachusetts, Oliver and Stephens (1977) were able to document 14 natural and human-caused disturbances of varying magnitudes that had occurred at irregular intervals between 1803 and 1952, and there was evidence of two hurricanes and a fire prior to 1803. Small disturbances did not bring in new species of trees, but they often allowed species already present in the understory, such as black birch, red maple, and hemlock, to emerge to the canopy. Large-scale disturbances, such as hurricanes or large fires, created openings into which early successional species, such as white birch or pin cherry, invaded. Oliver and Stephens concluded from their study that the present composition of the forest is more the result of allogenic forces than of autogenic development, or, to put it another way, that the present-day forest is a mixture of mature, early seral, and disturbance-modified vegetation. When development is "set back" by a disturbance such as a storm, a fire, or intense grazing, recovery may follow a different path in regard to species replacement than the original or primary succession, a pattern that has been called deflected succession (Goodwin 1929; Gibson and Brown 1992).

Dynamic Beaches

A sea beach is a good place to observe the interplay of auto- and allogenic processes. As long as wave action is gentle and the sand budget is balanced—that is, as much sand is deposited on average as is removed by tides and waves—the wind builds up sand dunes, and vegetation develops on them in an orderly sequence: beach grasses, then hardy forbs, then woody shrubs, then trees such as junipers, pines, and oaks. The community gradually stabilizes the dunes so that they are resistant to high tides and ordinary storms. If the sand budget becomes positive, the beach moves seaward and more dune succession occurs. However, if the sand budget becomes negative, perhaps because of a change in offshore currents, a rise in sea level, or the dredging and filling activities of humans, then the beach begins to shift landward, and the dunes may start to erode despite the vegetative cover. The dunes then become a source of sand to replenish and maintain the beach strand.

Only recently have scientists begun to understand this interaction of geophysical and biological forces. In the past, expensive seawalls, rock rip-raps, groins, and other artificial barriers were thought to be answers to beach erosion problems. In many cases these measures have not only

Where's the Beach?

On your next trip to the coast, notice that where there are seawalls (or wall-to-wall houses or hotels that block water flow) there is often no beach to walk on at high tide. Where economic investment in beach resorts is large, artificial beach nourishment, involving pumping or moving sand back onto the beach, may be justified. A more prudent approach to seashore development is to recognize the inherent instability of low-lying shores and to design human-made structures accordingly. For example, putting houses up on tall poles, so that high tides and storm waters can move freely under the structures and beyond, dissipates energy gradually and without harm, as on a natural beach. Many states and localities are now adopting building codes that require such prudent procedures.

proved futile, but have actually hastened the erosion of the beach. With a wall or other barrier in place, all of the energy of waves and tides is directed back onto the beach, washing it out and making the strand steeper, and the source of sand for natural repair is cut off by the obstruction. These tendencies are diagrammed in Figure 5. Thus, building a seawall may save your beach cottage (at least temporarily) but result in the loss of the beach—the reason you built (or bought) the cottage in the first place. For more on beach dynamics, see *The Beaches Are Moving* by Kaufman and Pilkey (1983).

Aging and Cyclic Succession

Even without external perturbations, the climax stage may not remain unchanged forever. Observations in old forests suggest that self-destructive biological changes, similar to what we would call aging in an individual, may be occurring. For example, young trees may not be quite replacing the old ones as they die, or regeneration of nutrients may be lagging and the community metabolism slowing down. There are so few

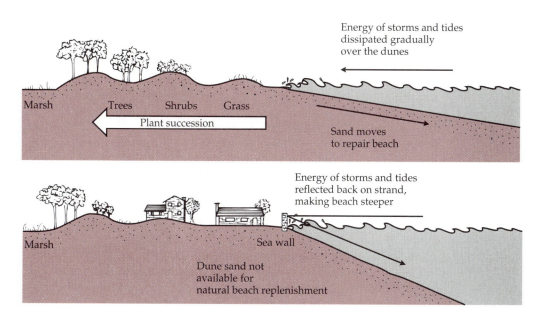

FIGURE 5
The effect of a sea wall or similar barrier on beach morphology—an interaction of auto- and allogenic processes.

quantitative studies at present that we cannot say whether natural communities suffer aging after reaching maturity (as do individual organisms), or whether, unlike organisms, unperturbed biotic communities are inherently capable of self-maintenance indefinitely.

Foresters commonly advocate the logging of old forests (which they class as "overaged") by arguing that since there is no longer any net growth, and since large trees are dying and rotting, the stand should be harvested or thinned so as to promote the growth of a new generation of young trees. While the forester may truly be thinking about the future, this recommendation is most likely motivated by the large amount of money one can get for the high-quality wood in large trees that have taken hundreds of years to grow. To be sure, the aesthetic, recreational, and life-support values of older forests are recognized by professional foresters as well as by landowners, but since these values are largely nonmarket, preservation loses out to logging when economics is the main consideration. On the other hand, landowners may opt for preservation if the land in question is not their sole source of income. Such is often the case in the vicinity of large cities, where urbanites own an increasing amount of rural land (Healy and Short 1981). There is a middle ground between economics and aesthetics, as we will note again in the next chapter.

The Georgia Piedmont forests and the city of Atlanta provide an example. As shown in Figure 1, the succession in these forests is from pines to hardwoods. Because pines have a higher market value than hardwoods, commercial forest management is directed toward arresting this successional trend so that the pine stages can be retained and regenerated. Urbanites, however, are often more interested in the recreational and second-home qualities of their land than in pulpwood production. Johnson and Sharpe (1976) report that despite silvicultural efforts to maintain pines, hardwoods increased in area in the Piedmont between 1961 and 1972, and they project, on the basis of a 30-year model, that hardwood stages will continue to increase, although at a slower rate than if only natural succession were involved. Urbanization and the suppression of fire, both of which favor hardwoods over pines, are important factors in the model. Thus, although the composition of the Piedmont forest is strongly influenced by the demand for pine products, its projected future composition will follow the trends of natural succession.

Questions of stability versus aging of mature systems are academic in many situations in which disease, storms, fire, and so on hasten the death of the community at or before climax and start a new cycle of seral stages. What the English ecologist A. S. Watt (1947) has called **cyclic succession** is a common pattern. The California chaparral, first mentioned in Chapter 2, is a good example. This dwarf woodland almost seems to program itself for periodic destruction by fire. As the community matures, litter and dead wood pile up faster than they can be decomposed during the long, dry summers. Antibiotic chemicals produced by the shrubs inhibit the growth of ground cover. The community becomes more and more combustible, and sooner or later, fire sweeps through the woodland. Accumulated litter is removed, antibiotics are neutralized, and shrubs and trees are killed back to ground level. Successional development then repeats itself: herbaceous vegetation develops from seeds and woody vegetation resprouts to grow to maturity again. In this way the aging community becomes youthful again for a while. What Sprugel and Bormann (1981) have called "wave-generated succession" in high-altitude forests is another example. As trees grow tall, they tend to be blown down and replaced by younger growth, so that a succession of young and mature vegetation moves as waves over the landscape in the general direction of prevailing winds.

The Pulse-Stabilized Subclimax

So far we have emphasized the destabilizing effects of allogenic physical surges. But acute perturbations can also be stabilizing if they occur in the form of regular pulses that can be utilized by adapted species as an extra energy subsidy. In fact, rhythmic, short-term perturbation imposed from without (as a forcing function, in model terminology) can maintain an ecosystem at some intermediate point in the successional sequence, resulting in a compromise between youth and maturity. What we call fluctuating water-level ecosystems are examples. Estuaries, intertidal shores, rice paddies, the Florida Everglades, and the New York Bight (described in Chapter 1) are held in highly productive early seral stages by the daily or seasonal rising and falling of water levels. The life cycles of the biota living in these systems are strongly adapted to the fluctuations. These pulse-stabilized subclimaxes (by subclimax we mean a developmental stage below, or short of, the climax that would develop in the absence of the perturbation) are very important components of the landscape because, like young systems, they have a high net production rate. Some of this production passes into and helps nourish neighboring systems, which may be less productive but may have redeeming aesthetic and life-support values and support valuable species not found in early successional stages (see the discussion of source–sink energetics in Chapter 4).

The Significance of Ecosystem Development to Land Use Planning

Nature's land use plan, which involves developmental trends toward increasing structure and complexity per unit of energy flow (high B/P efficiency; see Table 1), contrasts and often conflicts with the human economic goal of maximizing production, that is, trying to obtain from the landscape the highest yield of marketable products possible (high P/B efficiency). Recognizing the difference in "game plans" between the natural and the domesticated environments helps us understand the land use conflicts that increasingly arise as we strive to establish rational policies for managing the environment in our best long-term interest. We can think of natural succession as a "protective" strategy, since accumulated organic structures, stored nutrients, and diversity in the landscape are a hedge against unfavorable times, just as savings in the bank and equity in a home protect against economic hard times. In contrast, a "productive" strategy, as employed by humans, involves developing and maintaining early successional types of ecosystems that produce food, fiber, and other products that are harvested at the peak of growth, leaving little to protect the landscape from weather, wastes, and other physical uncertainties. But humans do not live by food and fiber alone. We also need a balanced carbon dioxide/oxygen atmosphere, the climate buffer provided by oceans and masses of vegetation, and clean (that is, unproductive) water for cultural and industrial uses.

In a nutshell, we need both productive and protective ecosystems. The most pleasant and certainly the safest landscape to live in is one containing a mixture of communities of different ecological ages, that is, a variety of crops, forests, lakes, streams, vegetated roadsides, marshes, seashores, and "waste places." (Incidentally, you will note in leafing through a wildflower guide that many interesting and beautiful flowers are found in those "waste places" that somebody—a highway engineer, for example—is always trying to replace with mowed grass or some other neat but dull ground cover.) As pointed out many times in this book, large areas of both natural and domesticated landscapes are necessary to support our high-energy, highly developed, but ecologically parasitic urban districts.

Since it is impossible to maximize conflicting uses in the same system at the same time, two solutions to "having our cake and eating some of it too" suggest themselves. We can compromise between yield and quality of environment, or we can deliberately compartmentalize the landscape (hopefully with great wisdom and foresight) so as to maintain both highly productive and predominantly protective types as separate units subject to different management strategies, ranging, for example, from intensive cropping to wilderness management. Conservation tillage agriculture, first mentioned in Chapter 5, is a good example of a successful compromise strategy, since good crop yields and the quality of the soils are both maintained, and chemical pollution of adjacent life-giving waters is reduced. Desirable compartmentalization is achieved when we set aside

A Vacation in a Cornfield?

Many essential life-support resources, not to mention aesthetic and recreational needs, are best provided by the parts of the landscape that exhibit low net production. In other words, the landscape is not just a supply depot, it is also the *oikos*—the home—in which we live. We can all admire a vast Iowa cornfield and appreciate its importance, but few of us would want to live in the middle of it, and we most certainly would not go there for recreation! As individuals we more or less instinctively surround our houses with protective, nonedible cover (trees, shrubs, grass) at the same time we strive to coax extra bushels from our cornfields. And for vacations, national parks and other natural areas are much more popular than croplands.

parks and other not-to-be-developed buffers and green spaces, and when we enact appropriate zoning ordinances, conservation easements, and open space legislation.

In the United States, as elsewhere, the amount of natural environment set aside as parks, national and state forests, wildlife refuges, wilderness areas, and so on varies from less than 10 percent of the land area in many eastern and midwestern states to more than 50 percent in some western states. A minimum of 20 percent (more in dry climates such as the western United States) would seem to be a reasonable planning goal in view of the demand for outdoor recreation and the need of an increasingly urban civilization for life-support environment. Ideally, every child should be able to walk or bicycle to a natural area to explore the secrets of nature.

Network Complexity Theory

The shift in energy use from growth to maintenance that we have cited as possibly the most important trend in ecological succession (see the energy flow model in Figure 3) has its parallel in growing cities and countries. People and governments consistently fail to anticipate that as population density increases and urban–industrial development intensifies, more and more energy, money, management effort, and tax revenues must be devoted to the services (e.g., water, sewage, transportation, and police) that maintain what is already developed and "pump out the disorder" inherent in any complex, high-energy system. Accordingly, less energy is available for new growth, which eventually can come only at the expense of the development that already exists. C. S. Shannon, the "father of information theory," puts it this way: Increasing disorder is a property of all complex systems. What has come to be know as the "network law" can be stated as follows:

$$C = \frac{N(N-1)}{2}$$

or approximately $N^2/2$.

In other words, the cost (C) of supporting a network (N) of services is a power function—roughly, a square—of N. That is, when a city or development doubles in size, the cost of maintenance may quadruple (see Pippenger 1978).

When Succession Fails

The title of this section is also the title of an essay by George M. Woodwell (1992), who writes succinctly and with great urgency about the environmental follies of humankind and the need to take action now to deal with global threats such as atmospheric toxification and global warming. Ordinarily when a landscape is devastated by storms, fires, or other periodic catastrophes, ecological succession is the healing process that restores the ecosystem. But when landscapes are severely abused over long periods of time (eroded, salinated, stripped of all vegetation, contaminated with toxic wastes, and so on), the land or water becomes so impoverished that succession cannot occur even after the abuse stops. Such sites represent a new class of environment that will remain barren indefinitely unless explicit efforts are made to restore it. Woodwell suggests that one-third of India is already in this degraded condition.

When ecosystem development fails, we have to resort to ecosystem *re*development. Perhaps this is why there are so many recent books, journals, and papers dealing with what is called restoration ecology (Cairns et al. 1971 and 1992; Jordan et al. 1987, to name three).

Evolution of the Biosphere

As is the case with short-term development, the long-term evolution of the biosphere is shaped by the interaction of geological and climatic forces with the autogenic processes resulting from the activities of living components. We have already briefly outlined the history of life on earth in connection with our discussion of the Gaia hypothesis in Chapter 3. The

Succession in Human Societies

Just as there are parallels between development at the organismal and community levels, there are interesting parallels between ecosystem development and the development of human societies (although these parallels are not based on the same cause-and-effect relationship). In pioneer societies, as in early stages of succession, opportunistic exploitation of the environment and accumulation of resources are necessary for the survival and growth of human populations. Accordingly, "clearing the land" is the first order of business. Exploitation of people (slavery) has also been common in the early stages of societal development. Protecting and maintaining life-support resources (and human rights, for that matter) does not receive a high priority at this stage, chiefly because the supply is still perceived to be larger than the demand. Ruins of civilizations and human-made deserts in various parts of the world stand as evidence that societies preoccupied with population growth and economic growth (and sometimes with war) often do not recognize the need for protective as well as productive environments until it is too late. After the hills are denuded and all the soil is washed off, only a massive infusion of outside capital can rehabilitate the land (as is now occurring to some extent in Israel). Likewise, when urban growth is allowed to mushroom unrestricted, it may be too late to protect air and water quality.

development of the biosphere in terms of the oxygenation of the atmosphere as linked with the evolution of the biota is shown in Figure 6.

Although we may never know exactly how life began on earth, the generally accepted theory is that the first living things were tiny anaerobic heterotrophs that lived on organic matter synthesized by abiotic processes. The composition of the atmosphere at that time was largely determined by the gaseous stuff that came out of volcanoes; the geologist would speak of this as atmospheric formation by crustal outgassing (Cloud 1988). This primordial atmosphere contained lots of nitrogen, hydrogen, carbon dioxide, and water vapor, and also contained carbon monoxide, chlorine, and hydrogen sulfide in quantities that would be poisonous to humans and to much of present-day life. The earth's early **reducing atmosphere** (a term used to contrast it with our present-day

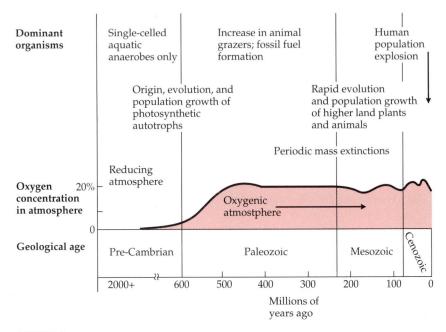

FIGURE 6
The evolution of the biosphere and its effect on the atmosphere.

oxygenic atmosphere) may have been similar to that now found on Venus or Jupiter. In the absence of gaseous oxygen, there was no protective ozone layer as there is now. Thus, at first, life could exist only if shielded by water or by some other barrier. But strange to say, it was the short-wave solar radiation that is now shielded out that is thought to have created the chemical evolution leading to the complex organic molecules that became the building blocks of (and also the food for) early life. According to this view, the conditions for creating life out of organic matter in that manner are no longer present on earth—and such conditions would be lethal to most present-day life!

Cyanobacteria: The First Great Breakthrough

For a billion or more years, life apparently maintained only a tiny foothold, limited in habitat and energy source, in a violent physical world. The big change began with the appearance at least two billion years ago of the first photosynthetic microorganisms, the **cyanobacteria** (also called blue-green algae, although they are not algae), which were

Our Atmosphere: A Drama

I can think of no better way to dramatize the absolute dependence of human beings on other organisms in the environment than to recount how our atmosphere came into being—emphasizing, of course, that it was built by microbes, not by humans (who are just now beginning to understand how it is maintained and how our massive energy transformations threaten its stability). The story of our air is a fascinating drama with enough mystery and potential tragedy to intrigue all readers. Berkner and Marshall have written both a literary account (1966) and a more technical treatise (1965), which provide good references, as do books by Cloud (1988), Margulis (1982), and Margulis and Sagan (1986).

able to use the sun's energy to make food from simple inorganic materials, and which released gaseous oxygen as a by-product. As the oxygen diffused into the atmosphere, the much more energy-efficient aerobic organisms evolved, and the ozone shield developed, making it possible for life to spread to all parts of the globe. There followed an almost explosive development of increasingly complex multicellular organisms. Over long stretches of time, production exceeded respiration ($P/R > 1$), so that oxygen increased and carbon dioxide decreased to present-day levels sometime in the Paleozoic age (Figure 6). Our fossil fuels were also formed during periods when P exceeded R by a wide margin and plant residues were fossilized in shallow seas and wetlands that covered a larger part of the earth than at present.

The cradle of life, of course, is still with us deep in the anaerobic sediments, and is well integrated with the vast aerobic world in which we oxygen-breathers live. Certain key roles that anaerobic processes play in the maintenance of biogeochemical cycles and atmospheric stability were described in Chapters 3 and 5.

Throughout geological history the diversity of life has increased despite periodic mass extinctions resulting from volcanic upheavals and comets striking the earth. After each such episode the number of individuals of each surviving species increases rapidly (in terms of geological time), but restoration of the diversity of species takes longer (Signor 1990). We see the same pattern in the short term when humans devastate a local habitat, whether they allow it to recover naturally or undertake active restoration measures.

The Mechanisms of Evolution

The word **evolution** (from the Latin, *evolutio*: unrolling) is widely used for temporal change, that is, change with time—usually with the connotation that the change is for the better (lower to higher, or from a simpler to a more complex state, for example). Thus, we may speak of the evolution of one's personality or the evolution of the automobile or airplane. **Organic evolution** refers to change in organisms with time, which, in general, involves long-term development from the simpler to the more complex or better-adapted condition. As outlined by Charles Darwin in his epoch-making treatise, *The Origin of Species* (1859), **natural selection** resulting from pressure by the environment and competing organisms is a major cause of change in organisms and species. Individuals that are best able to survive and produce the most offspring are thus "selected" by natural processes to populate the next generation. Darwin spoke of this as "survival of the fittest," and today the term "fitness" is used to indicate the ability of a species to persist to the next generation—an acceptable concept provided it is understood that the "fittest" are not necessarily the biggest and strongest, or the best fighters or competitors; many species have survived over the ages by more subtle means, such as camouflage or cooperation.

Darwin was well aware of nonviolent natural selection, but it was Peter Kropotkin who wrote in 1902 about the importance of mutual aid (mutualism) as a factor in evolution (see Chapter 6). Kropotkin also suggested that there were two kinds of natural selection, or two aspects to the struggle for existence, namely, organism versus organism, which leads to competition, and organism versus environment, which leads to mutualism. To survive, an organism does not compete with its environment as it might with another organism, but must adapt to or modify its environment and its community in a cooperative manner.

As knowledge of genetics advanced, it became evident that changing gene frequencies resulting from recurrent **mutations** (changes in individual genes) and from **genetic drift** (random changes in gene frequencies) provide the variation on which natural selection acts. The importance of genetic drift in evolution was first pointed out by the eminent geneticist Sewall Wright (1938). Genetic drift is a factor in the **genetic** or **population bottleneck** that may lead to extinction when the size of a population becomes very small.

Macro- versus Microevolution

Organic evolution is accepted by scientists as fact, not theory (although there is uncertainty and debate about the mechanisms involved). Ever

How to Survive and How to Become Extinct

At the risk of oversimplification, we might say that there are two contrasting ways to survive as a species: (1) invest a lot of energy in reproduction and rapid growth, thereby outpacing competitors, predators, and/or disease; or instead (2) invest a lot of energy in physical and/or chemical defense. In other words, these are the contrasting *r*- and *K*-strategies, one appropriate for pioneer situations and the other for mature, crowded communities.

Trees provide good examples of these strategies. Loehle (1988) has shown that for the most part trees fall into two distinct groups: (1) those, such as early successional species, that colonize quickly and grow rapidly but are short-lived and produce low-quality wood; and (2) those, such as oaks and hickories, that invest in defense, grow slowly, are long-lived, and produce dense, better-quality wood.

Again at the risk of oversimplification, we might say that there are two ways for a species to become extinct: (1) allow mortality to exceed reproduction (deterministic); or (2) become a victim of large-scale catastrophe (stochastic), as has happened to species in past geological ages. Human beings are currently in no danger of becoming extinct due to lack of reproduction, but are we vulnerable to catastrophe—perhaps a catastrophe resulting from too much reproduction?

since Darwin, biologists have generally adhered to the notion that evolutionary change is a slow, gradual process involving many small mutations and changes in gene structure accompanied by continuous natural selection of those genetic changes that have survival value for individuals. However, gaps in the fossil record and failure to find transitional forms ("missing links") have led many paleontologists to believe in what Gould and Eldredge (1977) have called **punctuated equilibrium**. According to this theory, species remain unchanged in a sort of evolutionary equilibrium for long periods; then, once in a while, the equilibrium is "punctuated" when a small population splits off and rapidly evolves into an entirely different species without transitional forms being deposited in the fossil record.

So far, no one has come up with a good explanation of what might cause such "macroevolutionary" leaps. For fascinating essays on Darwin, evolutionary thought, form and function, earth's history, natural history, science, racism, and other interesting topics, see Stephen Jay Gould's *Ever Since Darwin* (1977), and his essay "Darwinism Defined: The Difference between Fact and Theory" in *Discover* magazine (1987).

Speciation

Speciation, the formation of new species and the development of species diversity, occurs when gene flow within a common pool is interrupted by an isolating mechanism. When isolation occurs through geographic separation of populations descended from a common ancestor, **allopatric** ("different fatherland") **speciation** may result. When isolation occurs through ecological or genetic means within the same area, **sympatric** ("joint fatherland") **speciation** is a possibility.

Allopatric speciation has been generally assumed to be the primary mechanism by which species arise. A classic example is the case of Darwin's finches, first described by Charles Darwin when he visited the Galápagos Islands off the coast of Ecuador during the famous voyage of the *Beagle*. From a common ancestor, several species evolved in isolation on the different islands and **adaptively radiated**—that is, they changed to take advantage of the variety of habitats and niches present on the islands. As shown in Figure 7, species present today include slender-billed insect eaters, ground- and tree-feeders, large- and small-bodied finches, and even a woodpecker-like finch that uses thorns as tools to dig out insects in bark.

Jerky Evolution

The fossil record of the Bryozoa (small, plantlike, mostly marine animals that encrust rocks and other underwater surfaces) illustrates how "jerky" or "punctuated" speciation can be. As documented by Jackson and Cheetham (1994), new species of these animals appeared very abruptly after long periods of evolutionary stasis (that is, periods of little or no change). What could be the reason for this pattern? One possible explanation is that the small random mutations that occur all the time do not increase fitness; only large macromutations, which occur infrequently, have survival value and therefore result in the formation of new species (Santiago et al. 1996).

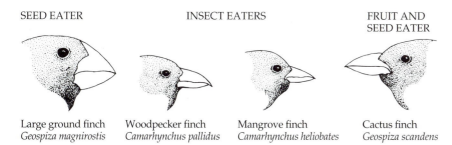

SEED EATER   INSECT EATERS   FRUIT AND
                                                                        SEED EATER

Large ground finch   Woodpecker finch   Mangrove finch   Cactus finch
Geospiza magnirostis *Camarhynchus pallidus*   *Camarhynchus heliobates* *Geospiza scandens*

FIGURE 7
Darwin's finches. Each of the species pictured has a distinctive bill shape
suited to its diet. (After Bowman 1961.)

You can visit the Galápagos on guided tours (as do thousands of tourists)
and see for yourself, or you can read about these birds in David Lack's
1947 book, *Darwin's Finches*, or in the more recent treatise by Grant (1986).

Isolation and subsequent speciation of terrestrial and freshwater
organisms was affected on a major scale by what has come to be known
as **continental drift**, the separation of the present-day continents from a
large common land mass. Sometime in the late Paleozoic era the northern
continents (North America, Eurasia) began to separate; then in the Meso-
zoic era, South America and Africa split and Australia separated from
Asia. The theory of continental drift is now accepted by most geologists
and is supported by the fossil record (Kurtén 1969; J. T. Wilson 1972; Van
Audel, 1995).

Evidence is mounting that strict geographic separation is not necessary
for speciation, and that sympatric speciation may be more widespread
than previously believed. Populations can become genetically isolated
from each other within the same area as a result of changes such as colonization, restricted dispersal of propagules, asexual reproduction, predation, and the like. In time, sufficient genetic differences accumulate in a
local area to prevent interbreeding.

Artificial Selection and Genetic Engineering

Selection carried out by humans to adapt plants and animals to their
own needs is known as **artificial selection** (in contrast with natural selection). Domestication, using this term in the broadest sense to include
both the cultivation of plants and the domestication of animals, involves
more than modifying the genetics of species, because reciprocal adaptations between the domesticated and the domesticator are required. We

are just as dependent on the corn plant, for example, as the corn plant is on us. A society that depends on corn develops a very different culture than one dependent on herding cattle. Domestication leads to a special form of mutualism and creates a special type of landscape, as described in Chapter 1.

Breakthrough discoveries about the biochemical nature of the genetic material (DNA) and the development of techniques for adding, removing, and modifying genes at the cellular level—**gene splicing**—promise to greatly speed up artificial selection. What is now known as **genetic engineering** or **biotechnology** is developing into a major technological revolution. If we judge by what happened during previous technological revolutions (for example, the development of atomic energy, discussed in Chapter 4, and the development of chemical pesticides) we can expect benefits from this, but also costs and problems, especially if the technology is promoted too rapidly, before its total impacts are understood. Expectations and benefits are usually overestimated while costs and problems are underestimated or not anticipated during the early stages of the development of a new technology. So far, the production of medicines such as insulin by genetically engineered microorganisms has been one useful result of biotechnology.

The possibility that altered organisms released or escaped into the environment might become pests is a major concern for the future. As explained in previous chapters, many of the worst diseases and pests we now contend with are naturally occurring organisms introduced into new environments. Accordingly, it seems prudent that full-scale testing under conditions that simulate the open systems of nature be required before release of genetically altered organisms is permitted. There are many legal and ethical questions that have to be considered.

One of the first controversies over the proposed release of a genetically altered organism involved a new form of the ice-nucleating bacterium *Pseudomonas syringae*, which enhances frost damage to fruit and vegetable crops in California. By removing the gene that controls the production of the lipoprotein coats of these microorganisms, their ice-nucleating capacity can be eliminated. The release of large numbers of the altered bacteria (called *Pseudomonas minus*) onto the leaves of strawberry plants was proposed in the hope that they would replace the natural form (*Pseudomonas plus*) long enough to reduce frost damage during the frost season. This seemed like a reasonable experiment with little potential for environmental damage—until it was discovered that *Pseudomonas syringae* are not just "pests," but may have the redeeming feature of enhancing rainfall. It seems that their lipoprotein coats, when shed and wafted up into the clouds, form ideal nuclei for the ice formation that is necessary for rain to fall—and reduction of rainfall could be a lot worse than frost damage. This example illustrates the need to consider secondary as well as primary effects of any proposed release (Odum 1985).

Hopefully, both the promise and the problems of biotechnology will be better anticipated and dealt with than has been the case with atomic energy. In both cases environmental considerations will largely determine success and failure. How far we can, or will, go in actually creating new life, and thus in directing evolution, remains to be seen.

Island Biogeography

Islands provide natural laboratories for studying evolution, and as such have fascinated biologists and ecologists ever since Darwin's visit to the Galápagos. The interplay of isolation, immigration, and extinction on islands has attracted attention since MacArthur and Wilson published their theory of **island biogeography** (1967). Simply stated, the theory holds that the number of species and the species composition of an island is dynamic (constantly changing) and is determined by the equilibrium between immigration of new species and extinction of those already present. Since numbers of species increase and decrease in an approximate logarithmic manner, and since rates of immigration and extinction depend on the size of an island and its distance from a mainland species reservoir, a general equilibrium model can be diagrammed, as shown in Figure 8. Thus, the smaller the island and/or the farther it is from shore, the fewer the species and the more unstable the species composition.

Schoener and Spillar (1995) carried out an interesting experiment in which spiders were introduced on islands in the Bermuda chain. They selected five islands in each of three categories: large with lizard predators, large without predators, and small without predators. Species of spiders were introduced that occurred elsewhere in the region, but not on the islands selected for the experiment. The introduced spiders survived best on large islands without lizard predators. Thus, both island size and predators were important in determining the success of colonization.

The theory of island biogeography, together with metapopulation theory, as outlined in Chapter 6, provides some useful guidelines for landscape planning and the establishment of nature preserves. Natural areas are increasingly pressured by **fragmentation**, in which natural habitats becomes "ecological islands" in a domesticated or urbanized landscape (see Figure 7 in Chapter 3). A large reserve may be preferable to a group of smaller ones of the same total area, since the larger the preserve, the greater the biotic diversity that can be supported. If one must settle for small parks, they should be close together or connected by corridors to

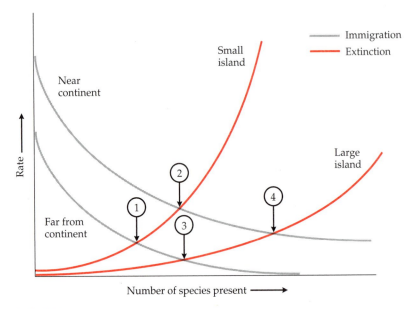

FIGURE 8
Theory of island biogeography. The number of species on an island is determined by equilibrium between immigration and extinction. Four equilibrium points are shown representing different combinations of large and small islands near and far from continental shores. (After MacArthur and Wilson 1963.)

facilitate continuous exchange of organisms. In the United States, establishment of a network of "wildlife corridors" between existing and proposed preserved natural areas is now being actively promoted at the state and regional levels both by private conservation groups, such as the Nature Conservancy, and by public agencies such as state departments of natural resources and the U.S. Fish and Wildlife Service (Harris 1984).

When a population becomes isolated on a small island-like patch of habitat, or otherwise becomes subdivided into small isolated colonies, genetic diversity becomes greatly reduced, and survival may depend on the presence of nearby larger colonies and the dispersal ability of the species in question. Sometimes restocking from larger, more diverse populations may be necessary to prevent extinction, as was discussed in Chapter 6. Vrijenhoek et al. (1985) describe such a case involving an endangered fish population. In another case, the effects of fragmentation due to logging in the Pacific Northwest have been extensively studied in hope of aiding the survival of the northern spotted owl (McKelvey et al. 1993).

Coevolution

Coevolution involves reciprocal natural selection between two or more groups of organisms with close ecological relationships but without exchange of genetic information between the groups (without interbreeding). Ehrlich and Raven (1965), who first proposed the term, used their studies of butterfly caterpillars and plants as a basis for proposing the hypothesis as follows. Plants, through occasional mutations or gene recombination, produce chemical compounds, perhaps as waste products, that are not harmful to the plant but turn out to be poisonous to an insect herbivore such as a caterpillar. Such a plant, now protected from the herbivore, thrives, and passes on the favorable mutation to successive generations. Insects, however, are quite capable of evolving strains tolerant to poisons—as is dramatically shown by the increasing number of insects that are becoming immune to insecticides. If a mutation or recombination appears in the insect population that allows individuals to feed on a previously protected plant, selection favors that genetic line. In other words, the plant and the herbivore evolve together, in the sense that the evolution of each depends on the evolution of the other. Pimentel (1968) has used the expression **genetic feedback** for this kind of evolution, which he demonstrated experimentally with flies and wasps, as described in Chapter 6.

Coevolution is presumed to have played a part in the origin of the mutualisms between plants and animals, animals and microbes, and so on, that were described in Chapter 6. However, it does not explain cooperation between species that are not closely linked by food or resource requirements.

Evolution of Cooperation and Complexity: Group Selection

To account for the incredible diversity and complexity of the biosphere and the widespread cooperation between species for mutual benefit, many scientists have postulated that natural selection operates beyond the individual and species level and beyond coevolution. **Group selection**, accordingly, is defined as natural selection between groups of organisms not necessarily linked by mutualistic associations. Group selection theoretically leads to the maintenance of traits that are favorable to populations and communities as a whole (the "public good") but that may be selectively disadvantageous to individuals within populations. D. S. Wilson (1980) states the case for group selection as follows:

Populations routinely evolve to stimulate or discourage other populations upon which their fitness depends. As such, over evolutionary time an organism's fitness is largely a reflection of its own effect on the community and the reaction of the community to that organism's presence. If this reaction is sufficiently strong, only organisms with a positive effect on their community persist.

How cooperation and elaborate mutualistic relationships get started and become genetically fixed is difficult to explain in evolutionary theory, because when individuals first interact, it is nearly always advantageous for each individual to act in its own interest rather than to cooperate. For example, when a fungus meets a tree root, it is advantageous for the fungus to try to feed on the root, and advantageous for the tree to repel the fungus. How, then, did mycorrhizal systems, in which the two organisms cooperate with mutual benefits, evolve? Axelrod and Hamilton (1981) and Axelrod (1984) have suggested a means by which reciprocation can develop as an extension of the conventional competition-based survival-of-the-fittest theory. They developed a model based on the "prisoner's dilemma game," in which two "players" decide whether or not to cooperate on the basis of immediate benefits. On first encounter, a selfish decision not to cooperate yields the highest rewards for each individual

When to Confront and When to Cooperate

We might suggest that the recent dramatic shift in the relationships between the superpowers is a parallel to the natural development of mutualism when things get tough. For several decades the United States and the Soviet Union increased their production of weapons in the name of defense. As the expense of this confrontation reached 15 to 20 percent of the wealth of each country, and as the defense initiative became a powerful political means to unite voters and win elections or maintain dictators, consumer, social, and environmental needs were neglected. As the pressure to address these internal needs became increasingly urgent, opportunities to move from Cold War to cooperation were eagerly seized by both countries. As we are finding out, this transition is difficult and will take time, but it is preferable to nuclear war. In other words, there are times and places for both confrontation and cooperation. A certain amount of confrontation is necessary to maintain order and to protect territories and cultures, but more important is the peace and cooperation needed to maintain human quality of life (Odum 1992).

regardless of what the other individual does. However, if both choose not to cooperate, they both do worse than if both had cooperated. If individuals continue to interact (i.e., the "game" continues), the probability is that cooperation may be selected on a trial basis, or it may simply occur on a random basis. If the cooperation is advantageous to both, then a partnership becomes established by natural selection acting at the individual level.

Both Wilson and Axelrod draw the analogy between the paradox of individual versus community fitness in biological communities and private versus public good in human communities. Axelrod further suggests that it is time for nations to shift from the competitive to the cooperative mode. Allman (1984) has written a provocative article based on Axelrod's work and ideas entitled "Nice Guys Finish First." Although few students of evolution doubt that group selection occurs, its importance in evolutionary history remains controversial. But then, although there is no doubt that organic evolution in general has occurred and continues to occur, we really do not yet understand all the intricacies of how it occurs. It would seem that long-term evolution has both individualistic and holistic components, as does short-term ecological succession. In other words, natural selection may occur at more than one level. Coevolution, group selection, and traditional Darwinism are all part of the hierarchical theory of evolution (Gould 1982).

Suggested Readings

*Allman, W. F. 1984. Nice guys finish first. *Science 84* 5(8): 24–32. (A review of Axelrod's book *The Evolution of Cooperation*. Theme: Cooperation pays in nature and society.)

*Axelrod, R. 1984. *The Evolution of Cooperation*. Basic Books, New York.

*Axelrod, R., and W. D. Hamilton. 1981. The evolution of cooperation. *Science* 211: 1390–1396.

*Berkner, L. V., and L. C. Marshall. 1965. History of major atmospheric components. *Proc. Natl. Acad. Sci. USA* 53: 1215–1226. (Literary account in the Tenth Anniversary Issue of *Saturday Review*, May 7, 1966, pp. 30–33.)

*Beyers, R. J., and H. T. Odum. 1993. *Ecological Microcosms*. Springer-Verlag, New York.

*Bowman, R. I. 1961. Morphological differentiation and adaptation in the Galapagos finches. *Occ. Pap. Calif. Acad. Sci.* 58: 1–302.

*Brooks, D. R., and E. O. Wiley. 1986. *Evolution as Entropy*. University of Chicago Press, Chicago. (Outcome of the constraints of the second law of thermodynamics is self-organization with the result that living systems exhibit increasing complexity and self-organization because of, not in spite of or at the expense of, entropy.)

*Cairns, J., ed. 1992. *Restoration of Aquatic Ecosystems*. National Academy Press, Washington, D.C.

*Cairns, J., K. L. Dickson, and E. E. Herricks, eds. 1977. *Recovery and Restoration of Damaged Ecosystems*. University Press of Virginia, Charlottesville.

Carson, H. L. 1987. The process whereby species originate. *BioScience* 37: 715–720. (Excellent discussion of current views on speciation.)

*Clements, F. E. 1916. Plant succession: An analysis of the development of vegetation. Publication no. 242. Carnegie Institution of Washington. (Reprinted in book form in 1928 by Wilson, New York.)

*Clements, F. E., and V. E. Shelford. 1939. *Bio-ecology*. John Wiley, New York.

*Cloud, P. E. 1988. *Oasis in Space: Earth History from the Beginning*. Norton, New York.

Connell, J. H., and R. O. Slayter. 1977. Mechanisms of succession in natural communities and their role in community stability and organization. *Am. Nat.* 111: 1119–1144. (Reviews several theories of succession.)

*Cooke, G. D. 1967. The pattern of autotrophic succession in laboratory microecosystems. *BioScience* 17: 717–721.

*Cowles, H. C. 1899. The ecological relations of the vegetation of the sand dunes of Lake Michigan. *Bot. Gaz.* 27: 95–391. (The pioneer American study of ecological succession.)

*Darwin, C. 1859. *The Origin of Species*. John Murray, London. (Facsimile edition ed. by E. Mayr. 1964. Harvard University Press, Cambridge, MA.)

Dawkins, R., and J. R. Krebs. 1979. Arms race between and within species. *Proc. R. Soc. London B* 205: 489–511.

Dolan, R., P. J. Godfrey, and W. E. Odum. 1973. Man's impact on the barrier islands of North Carolina. *Am. Sci.* 61: 152–162. (Illustrated account of auto- and allogenic impact on beaches.)

*Ehrlich, P. R., and P. H. Raven. 1965. Butterflies and plants: A study of coevolution. *Evolution* 18: 586–608.

*Gibson, C. W. D., and U. K. Brown. 1992. Grazing and vegetation change: Deflected or modified succession? *J. Appl. Ecol* 29: 120–131.

*Gleason, H. A. 1926. The individualistic concept of plant association. *Bull. Torrey Bot. Club* 33: 7–20.

*Goodwin, H. 1929. The sub-climax and deflected succession. *J. Ecol.* 17: 144–147.

*Gorden, R. W., R. J. Beyers, E. P. Odum, and R. G. Eagon. 1969. Studies of a simple laboratory microecosystem. *Ecology* 50: 86–100.

Gosselink, J. G., E. P. Odum, and R. M. Pope. 1974. *The Value of the Tidal Marsh*. LSU-SG-74-03. Center for Wetland Resources, Louisiana State University, Baton Rouge.

*Gould, S. J. 1977. *Ever Since Darwin*. Norton, New York.

*Gould, S. J. 1982. Darwinism and the expansion of evolutionary theory. *Science* 216: 380–387.

*Gould, S. J. 1987. Darwinism defined: The difference between fact and theory. *Discover* 8(1): 64–70.

*Gould, S. J., and N. Eldredge. 1977. Punctuated equilibria: The tempo and mode of evolution reconsidered. *Paleobiology* 3: 115–151.

*Grant, P. R. 1986. *Ecology and Evolution of Darwin's Finches*. Princeton University Press, Princeton, NJ.

*Harris, L. D. 1984. *The Fragmented Forest: Island Biogeography Theory and Preservation of Biotic Diversity*. University of Chicago Press, Chicago.

*Healy, R. G., and J. L. Short. 1981. *The Market for Rural Land*. The Conservation Foundation, Washington, D.C.

*Jackson, J. B., and A. H. Cheetham. 1994. Phylogeny reconstruction and the tempo of speciation in Cherlostone Bryozoa. *Paleobiology* 20: 407–410.

*Johnson, W. C., and D. M. Sharpe. 1976. An analysis of forest dynamics in the north Georgia Piedmont. *For. Sci.* 22: 307–322.

*Johnston, D. W., and E. P. Odum. 1956. Breeding bird populations in relation to plant succession of the Piedmont of Georgia. *Ecology* 37: 50–62.

*Jordan, W. R., M. E. Gilpin, and J. D. Aber, eds. 1987. *Restoration Ecology*. Cambridge University Press, New York.

*Kaufman, W., and O. H. Pilkey, Jr. 1983. *The Beaches Are Moving*. Duke University Press, Durham, NC.

*Kropotkin, P. 1902. *Mutual Aid: A Factor of Evolution*. William Heinemann, London. (Reprinted in 1935, Extending Horizons Books, Boston.)

*Kurtén, B. 1969. Continental drift and evolution. *Sci. Am.* 220(3): 54–65.

*Lack, D. L. 1947. *Darwin's Finches*. Cambridge University Press, Cambridge.

*Loehle, C. 1988. Tree life history strategies: The role of defense. *Canad. J. For. Res.* 18: 209–222.

*MacArthur, R. H., and E. O. Wilson. 1967. *The Theory of Island Biogeography*. Princeton University Press, Princeton, NJ. (See also MacArthur and Wilson 1963, *Evolution* 17: 373–387.)

*Margalef, R. 1963. Succession of populations. *Adv. Front. Plant Sci.* (New Delhi, India) 2: 137–188.

*Margalef, R. 1968. *Perspectives in Ecological Theory*. University of Chicago Press, Chicago.

*Margulis, L. 1982. *Early Life*. Science Books International, Boston.

*Margulis, L., and D. Sagan. 1986. *Microcosmos: Four Billion Years of Microbial Evolution*. Simon and Schuster, New York.

*McKelvey, K., B. R. Noon, and R. H. Lamberson. 1993. Conservation planning for species occupying fragmented landscapes: The case of the northern spotted owl. In *Biotic Interactions and Global Change*, eds. P. Kareiva, J. Kingsolver, and R. Huey, pp. 424–450. Sinauer Associates, Sunderland, MA.

Odum, E. P. 1969. The strategy of ecosystem development. *Science* 164: 262–270.

*Odum, E. P. 1985. Biotechnology and the biosphere. *Science* 229: 1338.

*Odum, E. P. 1992. When to confront and when to cooperate. *Georgia Landscape 1992*, p. 4. School of Environmental Design, University of Georgia, Athens.

*Oliver, C. D., and E. P. Stephens. 1977. Reconstruction of a mixed-species forest in central New England. *Ecology* 58: 562–572.

*Olson, J. S. 1958. Rates of succession and soil changes on southern Lake Michigan sand dunes. *Bot. Gaz.* 119: 125–176.

*Pimentel, D. 1968. Population regulation and genetic feedback. *Science* 159: 1432–1437.

*Pippinger, N. 1978. Complexity theory. *Sci. Am.* 238(6): 114–124.

*Santiago, F. E., V. S. Casper, and R. E. Lenski. 1996. Punctuated evolution caused by selection of rare beneficial mutations. *Science* 272:1802–1804.

*Schoener, T. W., and D. A. Spillar. 1995. Effect of predators and area in invasion: An experiment with island spiders. *Science* 267: 1811–1813.

*Shugart, H. H. 1984. *A Theory of Forest Dynamics: The Ecological Implications of Forest Succession Models*. Springer-Verlag, New York.

*Signor, P. W. 1990. The geologic history of diversity. *Annu. Rev. Ecol. Syst.* 21: 509–539.

*Sousa, W. P. 1984. The role of disturbance in natural communities. *Annu. Rev. Ecol. Syst.* 15: 353–391.

*Sprugel, D. G., and F. H. Bormann. 1981. Natural disturbance and the steady state in high-altitude balsam fir forests. *Science* 211: 390–393.

Stearns, S. C. 1983. Rapid evolution in ecological time. *BioScience* 33: 460.

Stanley, S. M. 1987. Periodic mass extinctions of the earth's species. *Bull. Am. Acad. Arts Sci.* 40(8): 29–48. (Mass extinctions appear to be pulsed at intervals of 20 million years or so, suggesting extraterrestrial causes such as comets striking the earth. Ice ages had little effect, as ice sheets covered only a small part of the globe.)

Ulanowicz, R. E. 1980. An hypothesis on the development of natural communities. *J. Theor. Biol.* 85: 223–245.

Ulanowicz, R. E. 1986. *Growth and Development: Ecosystem Phenomenology.* Springer-Verlag, New York.

*Van Audel, T. H. 1985. *New Views of an Old Planet: Continental Drift and the History of the Earth.* Cambridge University Press, New York.

*Vrijenhoek, R. C., M. E. Douglas, and G. K. Meffe. 1985. Conservation genetics of endangered fish populations in Arizona. *Science* 229: 400–402.

*Warming, E. 1909. *Oecology of Plants.* Clarendon Press, Oxford, England. (Originally published in Danish in 1895.)

*Watt, A. S. 1947. Pattern and process in the plant community. *J. Ecol.* 35: 1–22.

Wilson, D. S. 1976. Evolution on the level of communities. *Science* 192: 1358–1360.

*Wilson, D. S. 1980. *The Natural Selection of Populations and Communities.* Benjamin Cummings, Menlo Park, CA.

*Wilson, J. T., ed. 1972. *Continents Adrift.* W. H. Freeman, San Francisco.

*Woodwell, G. W. 1992. When succession fails. In *Ecosystem Rehabilitation: Preamble to Sustainable Development*, vol. 1, ed. M. K. Wali, pp. 27–35. SPB Academic Publishers, The Hague.

Wright, S. 1938. Size of population and breeding structure in relation to evolution. *Science* 89: 430–431. (The "Sewall Wright effect" and the importance of genetic drift.)

*Indicates references cited in this chapter.

8

Major Ecosystem Types of the World

For the most part in this book, we have based our approach to ecology on the analysis of units of the landscape as ecological systems. Principles and common denominators that apply to any and all situations, whether aquatic or terrestrial, natural or human-made, have been emphasized. The importance of the natural environment as the life-support module for Planet Earth and of the driving force of energy have been stressed. In Chapter 6, another useful approach was introduced, that of concentrating study on populations, which are the vehicles for evolutionary change. Still another useful approach is geographic, involving the study of the earth forms, climates, and biotic communities that make up the biosphere. In this chapter, we list and briefly characterize the major ecological formations or easily recognized ecosystem types (Table 1), with an emphasis on the geographic and biological differences that underlie the remarkable diversity of life on earth. In this manner we hope to establish a global frame of reference for the Epilogue, which deals with humankind's new challenge to attack problems on a large scale.

We would do well to start our world tour with the ocean, the largest and most stable ecosystem. The ocean, presumably, was the first ecosystem, for life is now thought to have originated in the saltwater milieu.

TABLE 1   Major Ecosystem Types and Biomes of the Biosphere
Marine Ecosystems
Open ocean (pelagic)
Continental shelf waters (inshore water)
Upwelling regions (fertile areas with productive fisheries)
Deep sea (hydrothermal vents)
Estuaries (coastal bays, sounds, river mouths, salt marshes)
Freshwater Ecosystems
Lentic (standing water): lakes and ponds
Lotic (running water): rivers and streams
Wetlands: marshes and swamp forests
Terrestrial Biomes
Tundra: arctic and alpine
Boreal coniferous forests
Temperate deciduous forests
Temperate grassland
Tropical grassland and savanna
Chaparral: winter rain-summer drought regions
Desert: herbaceous and shrub
Semievergreen tropical forest: pronounced wet and dry seasons
Evergreen tropical rain forest
Domesticated Ecosystems
Agroecosystems
Plantation forest and agroforest systems
Rural techno-ecosystems (transportation corridors, small towns, industries)
Urban-industrial techno-ecosystems (metropolitan districts)

The Ocean

The major oceans (Atlantic, Pacific, Indian, Arctic, and Antarctic) and their connectors and extensions cover approximately 70 percent of the earth's surface. Physical factors dominate life in the ocean (Figure 1A). Waves, tides, currents, salinity, temperature, pressure, and light intensity largely determine the makeup of the biological communities that, in turn, have considerable influence on the composition of bottom sediments and gases in solution and in the atmosphere.

The study of the physics, chemistry, geology, and biology of the ocean are combined into a sort of "superscience" called **oceanography**, which is

(A)   (B)

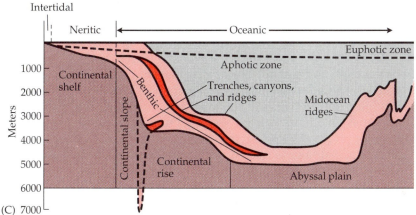

(C)

FIGURE 1
(A) The ocean. The never-ending wave motion seen in the photograph
serves to emphasize the dominance of physical factors in the open ocean.
(B) In many places the bottom of the ocean (in contrast to the surface) is
a relatively quiet and stable environment. The photograph shows an area
about 17 by 20 inches at a depth of 1500 meters on a transect between
Cape Cod and Bermuda in the Atlantic. Several brittle starfish are visible,
as well as worm tubes and two large worm burrows. (Photographs courtesy of Woods Hole Oceanographic Institution and D. M. Owen.) (C)
Zonation and bottom contour of the Atlantic. (After Heezen et al. 1959.)

becoming increasingly important as a basis for international cooperation.
Although exploration of the ocean is not quite as expensive as exploration
of outer space, a considerable outlay of money for ships, shore labora-
ries, equipment, and specialists is required. Most research is of necessity

carried out by a relatively few large institutions backed by governmental subsidies, mostly from the affluent nations.

The food chains of the ocean begin with the smallest known autotrophs and end with the largest of animals (e.g., giant fish, squid, and whales). Tiny green flagellates, algae, and bacteria that are too small to be captured by a plankton net are more important as a base for the ocean food web than the larger "net plankton" previously believed to fill that niche (Pomeroy 1974, 1984). Since a large portion of primary production is in dissolved form (DOM; see Chapter 4), DOM food chains are important in the open ocean (see Figure 10 in Chapter 4). A diversity of filter feeders, ranging from protozoa to pelagic mollusks that spin mucus nets to entrap microbes and detritus particles, provide the links between the small autotrophs and the large consumers.

To fully appreciate both the promise and the problems involved in human use of the ocean, we need to look at the contour of the ocean bottom, as shown in Figure 1C, which also gives standard oceanographic nomenclature for the zones of the ocean. As a citizen, you will hear a lot about the **continental shelf**, the sloping plateau that borders the continents. Located here is the bulk of the undersea oil and mineral wealth, as well as most of the seafood we now harvest. From the edge of the shelf, which varies greatly in width from location to location, the **continental slope** drops off rapidly into the true depths of the ocean. The topography of the continental slope is rugged, with huge canyons and ridges that are constantly changing under the forces of volcanic action and underwater landslides.

According to the now widely accepted "continental drift theory," some of the continents, notably Africa and South America as one pair and Europe and North America as another, were once single land masses, but drifted apart through the ages. The **mid-ocean ridges** (Figure 1C) are, according to this theory, lines of former contact between continents now hundreds of miles apart. Along these ridges and elsewhere, spreading tectonic plates create vents, hot sulfurous springs, and seeps. These **hydrothermal vents** support unique geothermal-powered communities unlike anything else so far discovered in the ocean (Figure 2). Here the food web begins with chemosynthetic bacteria rather than photosynthetic organisms. These bacteria obtain their energy to fix carbon and produce organic matter by oxidizing hydrogen sulfide (H_2S) and other chemicals. Filter- and suspension-feeding animals consume these bacteria in the plumes of hot water; snails and other grazers feed on the bacterial mats in the vent structures; and large tubeworms and clams have evolved a mutualistic relationship with some chemosynthetic bacteria, which live in the animals' tissues. And there are fish and crab predators. Most of the 300 or so species that have been discovered since the first vent community was

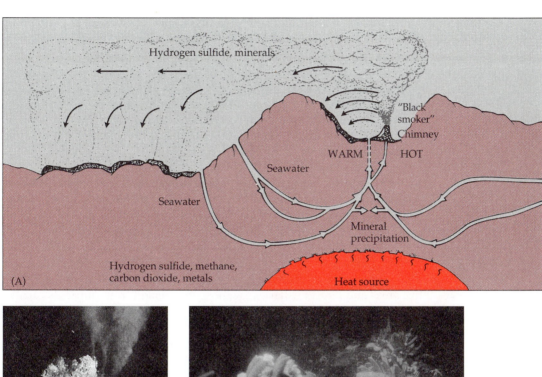

Hydrogen sulfide, minerals

"Black smoker" Chimney

WARM   HOT

Seawater

Seawater

Mineral precipitation

Hydrogen sulfide, methane, carbon dioxide, metals

Heat source

(A)

(B)

(C)

FIGURE 2

(A) Diagram of a thermal vent. Volcanic vents in the mid-ocean ridges produce plumes of "black smoke" rich in hydrogen sulfide and minerals. The vent community extends to areas where precipitate from the smoke has enriched the ocean floor. (B) At a depth of 1500 meters, a "chimney" emits a sulfide-rich plume into water that reaches 335°C. The chimney is covered with limpets and undersea worms (vestimentiferans and polychaetes). (C) In another part of a thermal vent colony—the water temperature is now only 15°C—large tubeworms (*Riftia pachyptila*) dominate the ecosystem. Young tubeworms have guts and feed on plankton; adults lose their guts and depend for their food on symbiotic sulfur bacteria living in their tissues. Crabs and mussels also make their home among the enriched deposits. (Photographs courtesy of V. Tunnicliffe.)

found in 1977 were previously unknown. This ecosystem has evolved in almost complete isolation from the rest of ocean life (Tunnicliffe 1992). Interestingly, fauna somewhat similar to that of vent communities have been found on the bodies of dead whales that lie on the ocean bottom (Smith et al. 1989). Like huge logs in the forest, this habitat fades away when the great carcasses are completely decomposed.

Since there are likely to be phytoplankton under every square meter of water, and since life in some form extends to the greatest depths, the oceans are the largest and "thickest" of ecosystems. They are also biologically very diverse, since many of the major taxonomic groups (phyla) are found only in the ocean. Marine organisms exhibit an incredible array of adaptations, ranging from the flotation devices that keep the tiny plankters within the upper layers of water, to the huge mouths and stomachs of deep-sea fish that live in a dark, cold world where meals are bulky but few and far between. As was shown in Figure 6 in Chapter 4, the continental shelf areas are fairly productive; seafood harvested here is an important source of protein and minerals for humans. The most productive areas and largest fisheries are those that benefit from nutrients carried up into the euphotic zone by currents, a process known as **upwelling**. Strong upwelling occurs in certain areas along the west coasts of several continents. The upwelling region along the coast of Peru is one of the most productive natural areas in the world. In contrast, vast stretches of the deep sea are mostly semidesert, with considerable total energy flow (because of the large area) but not much per unit of area. The autotrophic layer (photic zone) is so small in comparison with the heterotrophic layer (aphotic zone) that the nutrient supply in the former is easily depleted. In addition to tapping the potential energy of vertical temperature differences to generate electricity (as outlined in Chapter 4), we may be able to use some of this energy to culture seafood. Experiments with floating platforms or "reefs" in the photic zone on which seaweeds, crabs, and shellfish are cultured show some promise (Adey 1987). But even if we are never able to obtain much food from the deep sea, it is nevertheless very important to us. The oceans serve as giant regulators that help to moderate land climates and maintain favorable concentrations of carbon dioxide and oxygen in the atmosphere.

The polar and high mountain icecaps are extreme environments, but they are not lifeless. What Swan (1992) calls the Aeolian biome (from *aeolus*, "of the wind") is an ecosystem energetically based on ice algae and debris blown in from tundra and forest. Microorganisms and even a few insects can be found here. Humanity, of course, has a vital interest in what happens to icecaps, because if they melt, the sea will rise, to the detriment of perhaps half the world's population.

Mining the Ocean

International conferences are being held to discuss the thorny problem of setting up international law with rules and regulations for exploiting seabed minerals and energy resources. Most objective assessments (see, for example, Cloud 1969) warn against undue optimism that the deep sea is a vast storehouse waiting to be exploited. Recovering resources from the deep sea will be even more expensive than getting minerals and oil from the continental shelf, where costs are already huge. The sea is more important to us as a life-support system and climate regulator than it is as a supply depot. Anything we do to exploit the latter must not jeopardize the former.

Estuaries and Seashores

Between the oceans and the continents lies a band of diverse ecosystems. These are not just transition zones, but have ecological characteristics of their own (i.e., they are true ecotones; see Chapter 3). Although physical factors such as salinity and temperature are much more variable near shore than in the ocean itself, food is so plentiful there that the region is packed with life. Along the shore live thousands of adapted species that are not to be found in the open sea, on land, or in fresh water. Four kinds of marine inshore ecosystems—a rocky shore, a sandy beach, an intertidal mud flat, and a tidal estuary dominated by salt marshes—are shown in Figure 3 (following two pages).

FIGURE 3
Four types of coastal ecosystems. (A) A rocky shore on the California coast, characterized by underwater seaweed beds, tide pools with colorful invertebrates, sea lions ("seals"), and sea birds. (B) A sand beach with a ghost crab near its burrow. (C) An intertidal mud flat in Massachusetts during low tide. Although mud flats may look like deserts on the surface, they can, when not polluted or overexploited by humans, support very large populations of shellfish and other animals. (Photographs courtesy of U.S. Fish and Wildlife Service.) (D) A productive tidal estuary on the coast of Georgia, showing sounds, networks of tidal creeks, and vast areas of salt marsh. The shallow creeks and marshes not only support an abundance of stationary organisms, but also serve as nursery grounds for shrimp and fish that later move offshore, where they are harvested by trawlers. (Photograph courtesy of University of Georgia Marine Institute.)

(A)

(B)

(C)

(D)

The word **estuary** (from Latin *aestus*, tide) refers to a semienclosed body of water, such as a river mouth or coastal bay, where the salinity is intermediate between salt and fresh water, and where tidal action is an important physical regulator and energy subsidy. Estuaries and inshore marine waters are among the most naturally fertile in the world. Three major life forms of autotrophs are often intermixed in an estuary and play varying roles in maintaining a high gross production rate: phytoplankton; benthic microflora (algae living in and on mud, sand, rocks, and bodies or shells of animals); and macroflora (large attached plants, including seaweeds, submerged eelgrasses, emergent marsh grasses, and, in the tropics, mangrove trees). Estuaries provide the "nursery grounds" (i.e., place for young stages to grow rapidly) for most coastal shellfish and fish that are harvested not only in the estuary but offshore as well. Organisms have evolved many adaptations to cope with tidal cycles, thereby enabling them to exploit the many advantages of living in an estuary. Some animals, such as fiddler crabs, have internal biological clocks that help to time their feeding activities to the most favorable part of the tidal cycle. If such animals are experimentally removed to a constant environment, they continue to exhibit rhythmic activity synchronous with the tides.

An estuary is often an efficient nutrient trap that is partly physical (differences in salinity retard vertical but not horizontal mixing of water masses) and partly biological. This property enhances the estuary's capacity to absorb nutrients in wastes, provided organic matter has been reduced by secondary treatment. Estuaries have traditionally been much used, but little appreciated, as free sewers for coastal cities, as illustrated in our discussion of the New York Bight in Chapter 1. Since 1970, both awareness of and research on the values of estuaries has greatly increased. Most states have enacted legislation designed to protect these values.

Mangroves and Coral Reefs

Two very interesting and distinctive communities found in tropical and subtropical land–sea ecotones are mangroves and coral reefs. Both are potential "land builders" that help to form islands and extend seashores.

Mangroves are among the few emergent woody plants that tolerate the salinity of the open sea. A succession of species often forms a zone from open water to the upper intertidal region (Figure 4A). Extensive prop roots penetrate deeply into the anaerobic mud, bringing oxygen to the depths and providing surfaces for the attachment of clams, oysters, barnacles, and other marine animals. In Central America and Southeast Asia (Vietnam, for example) mangrove forests can have a biomass equal to that of a terrestrial forest. The wood is very hard and commercially valuable. In much of the tropics mangrove forests replace salt marshes as intertidal wetlands,

(A)

(B)

FIGURE 4

(A) Red mangrove (*Rhizophora mangle*) is usually the outermost species found in the mangrove forest ecotone. Its prop roots provide a substrate for oysters and numerous other marine organisms. Seeds sprout while still on the tree, then drop off and float in the water until they lodge in shallow water, perhaps to form a new island. (B) Underwater photograph of a coral reef, showing branching coral structures that develop where currents are gentle. More massive "heads" predominate on the exposed parts of coral reefs. (From *Fundamentals of Ecology* by E. P. Odum, 3rd Edition, 1971.)

and they have many of the same values, as, for example, serving as nursery grounds for fish and shrimp (W. E. Odum and McIvor 1990).

Coral reefs (Figure 4B) are widely distributed in warm, shallow waters. They form barriers along continents (for example, the Great Barrier Reef off Australia), fringing islands, and atolls (horseshoe-shaped ridges that develop on top of extinct underwater volcanoes). Coral reefs are among the most productive, diverse, and aesthetically pleasing of all biotic communities. No one should go through life without at least once donning a face mask and snorkel and exploring one of these colorful and prosperous natural cities.

Coral reefs can prosper in nutrient-poor waters because of water flow and a large investment in mutualism. Coral is a plant–animal "superorganism," since algae, called zooanthellae, grow in the tissues of the animal polyp. The animal component gets its "vegetables" from the algae growing in its belly, and obtains its "meat" by extending its tentacles at night to fish for zooplankton in the water flowing past its limestone house—which the colony builds by depositing calcium carbonate from raw material that is plentiful in the ocean. The plant component of this partnership gets protection and nitrogen and other nutrients from the animal component.

Because we humans must learn to prosper in a world of declining resources, the coral ecosystem serves as an example of how to efficiently retain, use, and recycle resources (Muscatine and Porter 1977). However, like a complex and energetic city, a finely tuned coral reef is neither resistant nor resilient to perturbations such as pollution or a rise in water temperature. In recent years the world's coral reefs have shown signs of stress that may be early warnings of global warming and oceanic pollution. An early sign of stress is the "bleaching" that occurs when the symbiotic algae leave the coral animal. If the mutualism is not restored, the coral slowly dies of starvation.

Streams and Rivers

Human history has often been shaped by the rivers that provide us with water, transportation, and a means of waste disposal (Figure 5A). Although the total surface area of rivers and streams is small compared with that of oceans and land masses, rivers are among the natural ecosystems most intensely used by humans. As in the case of estuaries, the need for multiple uses (as contrasted with a single-use approach to ecosystems such as cropland) demands that the various uses of rivers and streams (e.g., water supply, waste disposal, fish production, and flood control) be considered together and not as entirely separate problems.

The character of rivers changes from source to mouth. Not only do size and water volume increase, but community metabolism, species composition, and species diversity change as well. Stream ecologists speak of this longitudinal sequence as the **river continuum**. Upper tributary streams are often heterotrophic—that is, respiration exceeds production, with a P/R (photosynthesis: respiration) ratio of less than 1. The biotic community is largely dependent on organic matter washed in from the terrestrial watershed (or sometimes from adjacent lakes). In their middle sections, streams become wider and less shaded, and often become autotrophic (P/R of 1 or greater) as algae and other aquatic green plants become more numerous. Species diversity usually reaches a peak in the middle sections of rivers. In the lower reaches of large rivers, the current is reduced and the water is usually muddy, decreasing light penetration and aquatic photosynthesis. The stream then becomes heterotrophic again, with a reduced variety of species at most trophic levels.

Although streams are naturally adapted treatment systems for degradable wastes (recall the frequent comment about "free sewers"), almost all of the world's great rivers are severely overloaded with the residues of human civilization. In all parts of the world, rivers have been so extensively dammed, diked, and channelized that it is getting hard to find a truly wild river of any size—instead of a river continuum, we have created a river **serial discontinuum**. It is turning out that some of these manipulations bring only temporary or local benefits at great cost, and create additional problems costing still more money to correct (as in the case of some flood control projects). Accordingly, damaging floods that used to be "natural disasters" (and, therefore, unavoidable) are more and more proving to be human-made disasters (and, therefore, avoidable). Belt (1975) describes how such a situation has developed on the Mississippi River. Small floods are contained by dikes, but when water overtops the dikes in large floods, it spreads far beyond the floodplain, doing more damage than if the river had not been constrained in the first place. In the future, proposed alterations to rivers will have to be subjected to a more thorough cost–benefit analysis than was the case in the past.

The stream ecologist finds it convenient to consider flowing water ecosystems under two subdivisions: streams in which the basin is eroding and the bottom, therefore, is generally firm; and streams in which material is being deposited and the bottom, therefore, is generally composed of soft sediments. In many cases these situations alternate in the same stream, as may be seen in the rapids and pools of small streams. Aquatic communities are different in the two situations owing to the rather different conditions of existence. The communities of pools resemble those of ponds in

(A)

(B)

(C)

FIGURE 5
Freshwater ecosystems. (A) Convergence of two streams in northern New Jersey. The stream in the foreground flows from a watershed protected by grass and trees; the stream entering from the left is badly polluted with silt as a result of poor agricultural practices in its watershed. (Photograph courtesy of Soil Conservation Service.) (B) A natural pond in the grassland region of Western Canada. (C) A freshwater marsh in the Sacramento National Wildlife Refuge in California, where flocks of geese find refuge and shelter in productive aquatic and semiaquatic vegetation. (Photographs courtesy of U.S. Fish and Wildlife Service.)

that a considerable development of phytoplankton may occur, and the species of fish and aquatic insects are the same or similar to those found in ponds and lakes. The communities of the hard-bottom rapids, however, are composed of more unique and specialized forms, such as the net-spinning caddisworm (the larva of the caddisfly *Trichoptera*) which constructs a fine silk net that removes food particles from the flowing waters.

The load of sediments discharged into the oceans by the great rivers of the world tells us something about human abuses of the land. The rivers of Asia, the continent with the oldest civilizations and the most intense human pressure on the land, discharge 1500 tons of soil per square mile of land area drained by the rivers annually. In contrast, the sediment discharge rate for North America is 245, South America 160, and Europe 90 tons of soil per square mile of land area drained annually (data from Holeman 1968).

Lakes and Ponds

In the geological sense, most basins that now contain standing fresh water are relatively young. The life span of ponds (Figure 5B) ranges from a few weeks or months in the case of small seasonal ponds to several hundred years for larger ponds. Although a few lakes, such as Lake Baikal in Russia, are ancient, many large lakes date only as far back as the Pleistocene Ice Age. Standing-water ecosystems may be expected to change with time at rates more or less inversely proportional to their size and depth. Although the geographic discontinuity of fresh waters favors speciation, their lack of isolation in time does not. Generally speaking, species diversity is low in freshwater communities, and the same taxa (e.g., species, genera, and families) may be widely distributed over a continental mass and even between adjacent continents. A pond was considered in some detail in Chapter 3 as an example of a convenient-sized ecosystem for introducing the study of ecology.

Distinct zonation and stratification are characteristic features of lakes and large ponds. Typically, we may distinguish a **littoral zone** containing rooted vegetation along shore, a **limnetic zone** of open water dominated by plankton, and a deep-water **profundal zone** containing only heterotrophs. These zones parallel the major zones of the ocean shown in Figure 1C. In temperate regions, lakes often become thermally stratified during summer and again in winter, owing to differential heating and cooling. The warmer upper part of the lake (the **epilimnion**, from Greek *limnion*: lake) becomes temporarily isolated from the colder lower water (the **hypolimnion**) by a **thermocline** zone that acts as a barrier to the exchange of materials. Consequently, the supply of oxygen in the hypolimnion and of nutrients in the epilimnion may run short. During spring and autumn, as the entire body of water approaches the same temperature, mixing occurs. "Blooms" of phytoplankton often follow these seasonal rejuvenations of the ecosystem. In warm climates, mixing may occur only once a year (in winter), while in the tropics, mixing is a gradual or irregular process.

Primary production in a standing-water ecosystem depends on the chemical nature of the basin, the nature of imports from streams or land, and the depth. Shallow lakes are usually more fertile than deep ones, for the same reasons outlined in the discussion of the ocean. Accordingly, the yield of fish per acre of surface is generally inversely proportional to the mean depth. Lakes are often classified as **oligotrophic** ("few foods") or **eutrophic** ("good foods") depending on their productivity.

What has now come to be known as **artificial** or **cultural eutrophication** of lakes has created difficult problems in the vicinity of metropolitan areas and crowded resorts. Inorganic fertilizers in sewage effluent increase the primary production rates of lakes and change the composition of the aquatic community in ways that are not popular with the public. Game fish such as trout, which require cool, clear, oxygen-rich water, may disappear, and growth of algae and other aquatic plants may become so great as to interfere with swimming, boating, and sport fishing. Also, undecomposed dissolved organic matter may impart a bad taste to water even after it has passed through water purification systems. Thus, a biologically poor lake is preferable to a fertile one from the standpoint of water use and recreation. Again, we have a paradox—in some parts of the biosphere humans are doing everything possible to increase the fertility of water bodies in order to feed themselves, whereas in other places we do everything possible to prevent fertility (by removing nutrients, poisoning plants, and so on) in order to maintain a pleasant environment. A fertile green pond capable of producing many fish is not considered to be a good recreational swimming pool.

Efforts to divert municipal wastes from certain lakes have demonstrated that cultural eutrophication can be reversed in the sense that some lakes will return to a less fertile condition with improved water quality (in terms of human use) when nutrients no longer pour into them. Lake Washington in Seattle is a well-documented case (Edmondson 1968). In shallow-water lakes such as Lake Okeechobee in Florida, recovery is not so fast because wind-driven currents keep stirring up the bottom sediments, which continue to release the accumulated phosphorus and other nutrients.

Constructing artificial ponds and lakes (impoundments) is one of the conspicuous ways in which humans have changed the landscape in regions that lack natural bodies of water. In the United States almost every farm now has at least one farm pond, and large impoundments have been constructed on practically every river. Most of this activity works to benefit both humans and the landscape, for water and nutrient cycles are stabilized and the added diversity is a welcome change in our tendency to create a monotonous landscape. However, the impoundment idea can be carried too far; covering up fertile land with a body of water that cannot yield much food may not be the best land use.

People seem strangely unprepared for the changes that arise from ecological succession in artificial ponds and lakes. Somehow, once a lake has been created, it is expected to remain the same, as would a skyscraper or bridge. Instead, of course, all the processes of succession that were described in Chapter 7 take place, resulting from the activities of the biotic community (autogenic processes) and, especially in ponds and shallow lakes, from sediment discharges from the watershed (allogenic processes). Fishing is often good for the first few years in a new impoundment, then declines as the excess nutrients in the flooded watershed are exploited and the body of water begins to age, as shown in Figure 6.

Freshwater Marshes

Much of what was said about estuaries also applies to freshwater marshes (Figure 5C); which also tend to be naturally fertile ecosystems. Tidal action occurs in some coastal river marshes, and periodic fluctuation in water levels resulting from seasonal and annual rainfall variations often accomplishes the same thing in terms of maintaining long-range stability and fertility. Fires during dry periods consume accumulated organic matter, thereby deepening the water-holding basins and aiding subsequent aerobic decomposition and release of soluble nutrients, thus increasing the rate of production. In fact, if water-level fluctuations and fires do not occur, the buildup of sediments and peat (undecayed organic matter)

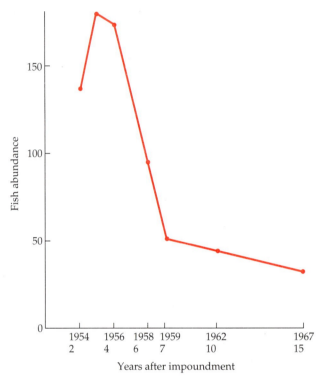

FIGURE 6

Fish abundance (based on mean of two sampling methods) in Lake Francis Case, South Dakota, a new mainstream reservoir on the upper Missouri River, from the second to the fifteenth year after completion of the dam and full impoundment of water. (Data from Gasaway 1970.)

tends to lead to the invasion of terrestrial woody vegetation. Where humans control water levels in marshes with dikes (for waterfowl conservation and hunting, for example), periodic drawdowns are required to slow the marsh-to-woody vegetation progression. Otherwise chemical herbicides or mechanical methods have to be used.

In tidal freshwater marshes along coastal plain rivers, the biota benefit from the tidal pulses but do not have to contend with salt stress. Accordingly, diversity of higher plants and animals is greater than in the tidal salt marshes downstream in the estuaries. As shown in Figures 7A and 7B, great masses of fleshy (low fiber content) vegetation are produced during the summer, then decompose back to mud level in winter, in contrast to the salt marshes (Figure 7C) where the more fibrous marsh grasses remain standing all year round. The anaerobic microbial populations are

Wetland Values

Wetlands are ecotones (see Chapter 3) that lie between land and water. Examples described or pictured in this chapter include salt marshes and mangroves that link land and sea; freshwater marshes; and riparian forested wetlands that surround rivers and lakes. There are also bogs, fens (a European term), and many other kinds of more isolated wetlands that are ecotones between land and groundwater. Wetlands are most productive when there is a strong energy convergence of sun and water flow ("naturally subsidized solar-powered ecosystems;" see Chapter 4). Accordingly, hydrology is the key to structure and function in wetlands.

Overall, wetlands deserve protection as buffers between human land development and our increasingly critical water resources. In addition to their role in maintaining water quality, wetlands have many other more specific values: as nursery grounds for fish and shellfish, habitat for water birds and other interesting and valuable wildlife, sources of valuable hardwood timber and other products, absorbers of flood waters, waste assimilators, and, on the global scale, as carbon sinks.

In the United States and a few other countries, coastal wetlands are legally protected. Interior wetlands, floodplains, and riparian zones, however, are not adequately protected from insidious development. Two things need to be recognized by public decision makers: (1) since wetlands are directly connected to rivers, lakes, estuaries, or groundwater, they are a part of our water resources, and hence part of the public domain—not real estate open to laissez-faire development; and (2) in the long term, wetlands are economically more valuable in their natural state than they are drained or otherwise converted to other uses (E. P. Odum 1988).

The most comprehensive reference to wetland values is a book edited by Greeson, Clark, and Clark (1979). See also the text by Mitsch and Gosselink (1994). The U.S. Fish and Wildlife Service has issued a series of "Community Profile" bulletins on various wetlands. (For a catalog, write National Wetland Center, 1010 Gause Boulevard, Slidell, LA 70458.)

dominated by sulfide reducers in salt marshes, whereas fermenters and methane producers are more numerous in freshwater marshes. In general, primary production in salt marshes culminates in aquatic animals—fish,

(A)

(B)

(C)

FIGURE 7
Fresh and saltwater tidal marshes. (A) A freshwater tidal river marsh in
the summer shows lush, fleshy vegetation. (B) During the winter the
marsh plants decompose back to mud level. (Photographs courtesy of
Carole McIvor.) (C) In contrast, the more fibrous marsh grasses of a tidal
saltwater marsh remain standing throughout the year. (Photograph courtesy of U.S. Forest Service.)

shrimp, and shellfish. Semiaquatic amphibians, reptiles, birds (ducks,
waders, etc.) and furbearers are more abundant in freshwater marshes.
(For further comparisons of these two marsh types, see W. E. Odum 1988).

The Florida Everglades is an exceptionally large and interesting stretch
of freshwater marshes characterized by naturally fluctuating water levels.
These great marshes are important in maintaining water tables under
coastal cities. Water levels tend to go down during the winter dry season
and rise during the summer wet season. In recent years dikes, drainage
canals, and water diversions north of the Everglades have so disrupted
this seasonal pulse that the wood stork, which breeds during the draw-
down when small fish are concentrated in isolated pools, has abandoned
the Everglades and moved north, where seasonal drawdowns are more
reliable. A lot of plans are being proposed to restore the Everglades as a

life-support module for the eastern Gold Coast of Florida (see, for example, Lodge 1994; Davis, Ogden, and Parks 1995).

It is significant that rice culture, one of the most productive and dependable of agricultural systems yet devised by humans, is actually a type of freshwater marsh ecosystem. The flooding, draining, and careful rebuilding of the rice paddy each year has much to do with the maintenance of the continuous fertility and high production of the rice plant, which itself is a kind of cultivated marsh grass.

Forested Wetlands

Swamp and floodplain forests occur in river bottoms, often intermixed with marshes, especially where large rivers cross coastal plains. They are also found in large depressions (the Okefenokee Swamp, for example; see Figure 8), lime sinks, and other low-lying areas that are flooded at least some of the time. As with marshes, hydrology plays a major role in determining species composition and productivity. Cypress and gum are the trees best adapted to flooding, while so-called bottomland hardwoods (lowland species of oak, ash, elm, and maple) do best where flooding is pulsed, as on floodplains. The greatest productivity occurs where the soil surface is flooded in winter or spring and is relatively dry during most of the growing season. A graphic model for the effect of flooding on wetlands in general is shown in Figure 9.

The Terrestrial Biomes

Large, easily recognized terrestrial community units are known as **biomes**. Throughout a given biome, the life form of the climax vegetation (see Chapter 7 for an explanation of the concept of climax) is uniform, and is the key to recognition of the biome. Thus, the dominant climax vegetation in the grassland biome is grass, although the species of dominant grasses vary throughout the different geographic regions where the grassland biome occurs. Other types of vegetation will be included in the biome, as, for example, "weedy" seral stages in succession, forest subclimaxes related to local soil and water conditions, crops, and other introduced vegetation.

The distribution of six major terrestrial biomes in relation to temperature and rainfall is shown in Figure 10. If you check the mean annual temperature and rainfall of your locality, you can determine from Figure 10 which biome you live in, even if you are sitting in the middle of a city with no climax vegetation anywhere around. Several other biomes (not shown in Figure 10), such as chaparral, tropical savanna, thorn shrub, and

(A)

(B)

FIGURE 8

A wetlands forest: the Okefenokee Swamp, Georgia. (A) Freshwater marsh is in the foreground, with the swamp forest in the background. (B) A cypress tree swells at the base and develops "cypress knees" (foreground) in response to frequent flooding.

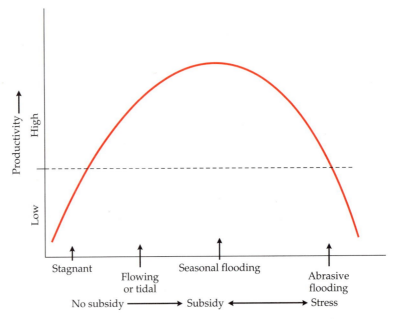

FIGURE 9
The relationship between productivity and a flooding gradient shown as a subsidy-stress model (see Figure 17 in Chapter 5). The dashed line represents the level of regional productivity, with moderate and seasonal flooding as a subsidy and stagnant or abrasive flooding as a stress.

tropical monsoon forest, are related to seasonal distribution of rainfall rather than annual mean rainfall.

Deserts

Desert biomes occur in regions with less than 10 inches (25 centimeters) of annual rainfall, and sometimes in hot regions where there is more rainfall, but it is unevenly distributed over the annual cycle. Lack of rain in the mid-latitudes is often due to stable high-pressure zones; deserts in temperate regions often lie in "rain shadows," that is, where high mountains block off moisture from the seas. Two types of North American deserts are shown in Figure 11, a "hot" desert in Arizona, where creosote bushes and cacti are conspicuous (Figure 11A), and a "cool" desert in Washington State, with sagebrush (Figure 11B). The characteristic spacing of desert vegetation and the possibility of the plants using repellent mechanisms were discussed in Chapter 6. North American deserts are

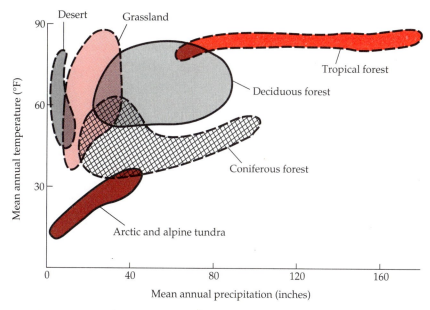

FIGURE 10
Distribution of the six major terrestrial biomes in terms of mean annual
temperature and mean annual rainfall in inches. (Courtesy of National
Science Foundation.)

not as extreme as some on other continents, such as the African Sahara or
the Asian Gobi. Some seasonal rain can be expected every year in deserts
in the United States, but rainless periods in extreme deserts may span
several years.

Four distinctive life forms of plants are adapted to the desert ecosystem. The annuals (such as cheat grass, shown in Figure 11B) avoid
drought by growing only when there is adequate moisture. Desert shrubs
have numerous branches arising from a short basal trunk and small, thick
leaves that may be shed during dry periods. They survive by their ability
to become dormant before wilting occurs. In cooler deserts, the shrubs
develop very long root systems that tap deep moisture that remains available after the surface completely dries out. In such cases the leaves and
stems may remain green and active throughout the summer. The succu-
lents, such as the cacti of the New World or the euphorbias of the Old
World, store water in their tissues. Microflora include mosses, lichens, and
cyanobacteria that remain dormant in the soil but are able to respond
quickly to cool or wet periods.

(A)

(B)

FIGURE 11

Two types of deserts in western North America. (A) A "hot" desert in Arizona. (Photograph courtesy of R. R. Humphries.) (B) A "cool" desert in eastern Washington State in early spring. (Photograph courtesy of Hanford Atomic Products Operation.) The desert shrub life form is illustrated by the dark creosote buses in (A) and the sagebrush in (B). Note the rather even spacing of the shrubs, especially evident in (A). The succulent life form is represented by cacti in (A) and the desert annual is represented by the cheat grass growing between the sage bushes in (B). (The objective of the radioactive tracer experiment shown in (B) was to determine the relative uptake from soil of specific minerals by the two life forms growing within the metal ring.)

Some reptiles and insects are "preadapted" to deserts because their waterproof body coverings and dry excretions enable them to get along on a small amount of water. Mammals as a group are poorly adapted to deserts, but a few species have become secondarily adapted. A few species of nocturnal rodents, for example, that excrete concentrated urine and do not use water for temperature regulation, can live in the desert without drinking water. Other animals, such as camels, must drink periodically, but are physiologically adapted to withstand tissue dehydration for periods of time. (For more on adaptations of desert animals, see Schmidt-Nielsen 1973.) In the past, remarkable human cultures have adapted to life in or along the edges of deserts, making use of desert-adapted plants and animals. Such a life in dry regions requires ingenuity and a conservation ethic, two attributes badly needed in more benign regions.

Because water is the dominant limiting factor, the productivity of a given desert region is an almost linear function of rainfall. In the California Mojave Desert, a 100-mm annual rainfall will result in about 600 kg/ha (kilograms per hectare) of net production, while 200 mm of rainfall will increase net production to about 1000 kg/ha. Where evaporative losses are less in the cooler Great Basin deserts, 200 mm of rainfall produces 1500–2000 kg/ha.

Where soils are suitable, irrigation can convert deserts into some of our most productive agricultural land. Whether productivity continues or is only a temporary "bloom" depends on how well humans are able to stabilize biogeochemical cycles and energy flow at the new increased rates. As a large volume of water passes through the irrigation system and evaporates from the soil, salts may be left behind that will gradually accumulate over the years until they become limiting—a process known as salinization, described in Chapter 5. The water supply itself can fail if the watershed from which it comes is abused. The ruins of old irrigation systems, and the civilizations they supported, in the deserts of the Old World warn us that the desert will not continue to bloom unless we understand that major alterations of the environment nearly always create problems as well as benefits, and that we must anticipate and deal with those problems before they become unmanageable.

Tundra

Between the forests to the south and the Arctic Ocean and polar icecaps to the north lies a circumpolar band of about 5 million acres of treeless country called the arctic tundra (Figure 12). Smaller but ecologically similar regions found above the tree line on high mountains are called alpine tundras. As in deserts, a limiting physical factor rules these lands, but it is heat rather than water that is in short supply in terms of biological func-

(A)

(B)

FIGURE 12

The tundra. (A) Closeup of tundra in August near the Arctic Research Laboratory at Point Barrow, Alaska, showing grass and sedge vegetation. (Photograph courtesy of R. E. Shanks and J. Koranda.) (B) Aerial view of tundra, showing herd of reindeer. The bumpy nature of the landscape is due to frost action; note also the numerous small ponds. (Photograph courtesy of U.S. Fish and Wildlife Service.)

tion. Precipitation is also low, but water as such is not limiting because of the low evaporation rate. Thus, we might think of the tundra as an arctic desert, but it can best be described as a wet arctic grassland or a cold marsh that is frozen for a portion of the year.

Although the tundra is often known as the "barren grounds" and might be expected to have a relatively low biological productivity, a surprisingly large number of species have evolved remarkable adaptations to survive the cold. The thin vegetation mantle is composed of lichens, grasses, and sedges, which are among the hardiest of land plants. During the long daylight (long photoperiod) of the brief summer, the primary production rate is high where topographic conditions are favorable (as in the low-lying areas of Figure 12A). The thousands of shallow ponds, as well as the adjacent Arctic Ocean, provide additional food to tundra food chains. There is enough combined aquatic and terrestrial net production, in fact, to support not only breeding migratory birds and emerging insects during the summer, but also permanent resident mammals that remain active throughout the year. The latter range from large animals such as musk oxen, reindeer (Figure 12B), polar bears, wolves, and marine mammals, to lemmings that tunnel about in the vegetation mantle. The dramatic ups and downs in the density of lemmings were discussed in Chapter 6. The large land herbivores are highly migratory, since there is not enough net production in any one local area to support them. Where humans try to "fence in" these animals or select nonmigratory strains for domestication, as with domestic reindeer, overgrazing is almost inevitable unless judicious rotation of pastures is employed to offset the absence of migratory behavior. The impact of humans on the tundra will increase as we strive to exploit oil and mineral resources from polar regions.

Grasslands

Natural grasslands occur where rainfall is intermediate between that of desert lands and forest lands. In the temperate zone, this generally means an annual precipitation of 10–30 inches (25–60 cm) depending on temperature, seasonal distribution of the rainfall, and the water-holding capacity of the soil. Tropical grasslands may receive up to 60 inches (120 cm) concentrated in a wet season that alternates with a prolonged dry season. Soil moisture is a key factor, especially as it limits microbial decomposition and recycling of nutrients. Large grassland areas occupy the interior of the North American and Eurasian continents. Other extensive natural grasslands are located in southern South America, central and southern Africa, and Australia.

Several aspects of North American grasslands are shown in Figure 13. The dominant plant life forms are the grasses, which range from tall species (5–8 feet) to short ones (6 inches or less) and may be bunch grass types (growing in clumps) or sod-formers (with underground rhizomes). A well-developed grassland community contains species with different temperature adaptations, one group (often C_3 plants) growing during the cool part of the season (spring and autumn) and another (often C_4 plants) during the hot part (summer); thus the grassland as a whole "compensates" for temperature, extending the period of primary production. The roles of C_3 and C_4 photosynthesis were discussed in Chapter 4. Forbs (nongrassy herbs) are often important components of grasslands, and woody plants (trees and shrubs) also occur, often in belts or groups along streams and rivers. Extensive areas in East Africa and other equatorial regions are occupied by a variant of the grassland biome, the tropical savanna, where trees with characteristic umbrella-like shapes are widely scattered over the grassland.

The grassland community builds an entirely different type of soil than a forest, even when both start with the same parent mineral material. Since grass plants are short-lived compared with trees, a larger amount of

(A)

FIGURE 13
Four aspects of grasslands. (A) Natural grassland with herd of bison on the National Bison Range in Montana. (Photograph courtesy of U.S. Fish and Wildlife Service.) (B) Cattle grazing in natural grassland that is in good condition. (C) Overgrazed grassland that has the appearance of a human-made desert. (Photographs courtesy of U.S. Forest Service.) (D) Grassland converted to intensive grain farming. (Photograph courtesy of U.S. Fish and Wildlife Service.)

(B)

(C)

(D)

organic matter is added to the soil. The first phase of decay is rapid, resulting in little litter but much humus; in other words, humification is rapid but mineralization is slow. Consequently, grassland soils may contain 5–10 times more humus than forest soils. The dark grassland soils are among those best suited for growing principal food plants such as corn and wheat (Figure 13D), which, of course, are species of cultivated grasses.

The role of fire in maintaining grassland vegetation in competition with woody vegetation in warm or moist regions was discussed in Chapter 5 (see Figure 12 in Chapter 5). Large herbivores are a characteristic feature of grasslands (Figure 13A). These are mostly large mammals, but large grazing birds are known to have occurred in the original fauna of New Zealand. The "ecological equivalence" of bison, antelope, and kangaroos in grasslands of different geographic regions was mentioned in Chapter 3. The large grazers occur in two life forms: running types, such as those mentioned above, and burrowing types, such as ground squirrels and gophers. When grasslands are used by humans as natural pastures, the native grazers are usually replaced with their domesticated kin—that is, cattle, sheep, and goats. Since grasslands are adapted to heavy energy flow along the grazing food chain, such a switch can be ecologically sound. Humans, however, have had a persistent history of misusing grassland resources by allowing overgrazing (Figure 13C) and overplowing. As a result, many grasslands are now human-made deserts. The importance of ecological indicators in the early detection of overgrazing was mentioned in Chapter 5.

Morello's (1970) outstanding study of the interaction of fire and cattle grazing in Argentina has traced how large areas of Argentina's grasslands have become covered with thorny shrubs. Morello showed that intensive cattle grazing reduces combustible matter so that the fires necessary to maintain grass cover can no longer burn. As a result, thorny shrubs, formerly kept in check by periodic fires, take over. The only way to restore grazing productivity is to expend fuel energy in mechanical removal and burning of woody vegetation. This is an example of a human-made vegetation change reversible only at great cost.

What to do about the African grasslands that contain an unusual diversity of mammalian grazers is a question now facing the emerging nations of that area as they strive to raise the nutritional levels of their expanding human populations. To accommodate the migratory behavior of herds, national parks or other protected areas must be large and/or connected by corridors; the animals cannot be saved by completely fencing them in. Some ecologists believe it may be feasible to harvest the antelope, hippopotamuses, and wildebeests for food on a sustained-yield basis rather than exterminating them in order to substitute cattle. For one thing, the natural diversity means broader use of primary production. Further, the native species are less vulnerable to the many tropical parasites and diseases that afflict cattle.

Forests

In Chapter 3, the point was made that the open sea and the forest are, in a comparative sense, the extreme natural ecosystem types in the biosphere in regard to standing crop biomass and the relative importance of allogenic and autogenic regulation. As shown in Figure 1 in Chapter 7, well-ordered and often lengthy ecological succession is characteristic in forest regions, with herbaceous plants often preceding trees. Consequently, in any one forest region we may see a mixture of vegetation, including nonforest stages in succession as well as forest variants that are adapted to special soil and moisture conditions.

Because the range of temperatures that will allow forest development is extremely wide, a sequence of forest types replace one another in a north–south gradient. Moisture is more critical to trees than to grasses, but forests occupy a fairly wide gradient from dry to extremely wet situations. Figure 14 shows three distinctly different forests in a north–south gradient. The northernmost forests (Figure 14A), which form a broad belt all the way around the top of the earth, are characterized by evergreen conifers of the genera *Picea* (spruce) and *Abies* (fir); species diversity is low, often with one or two species of trees forming pure stands. These circumpolar coniferous forests contain about a quarter of the earth's forest biomass and perhaps half of its forest carbon, so they are an important "sink" for excess carbon dioxide produced by human activity. It is from this belt that cold fronts sweep clean air across the densely populated temperate zones. Accordingly, these great forests are very valuable even if no timber is ever harvested from them.

Deciduous forests (Figure 14B) are characteristic of the more southern moist-temperate regions; these forests have more pronounced stratification and greater species diversity. These are the forests that are being extensively fragmented by urban and agricultural development (see Figure 7 in Chapter 3). Pines (*Pinus*) are found in both northern coniferous and temperate deciduous forest regions, often as seral stages.

Tropical forests (Figure 14C), the third forest type, range from broadleaved evergreen rain forests, where rainfall is abundant and distributed throughout the year, to tropical deciduous forests that lose their leaves during a dry season. Two life forms, the vines (lianas) and the epiphytes (air plants), are especially characteristic of tropical forests; a few

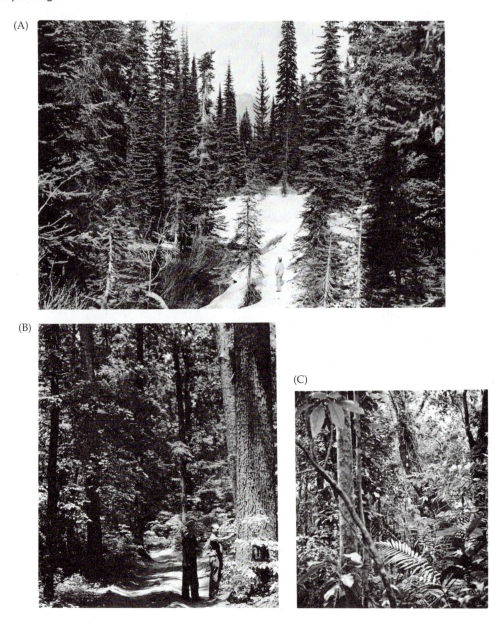

FIGURE 14
Three forest types in a north–south temperature gradient. (A) Northern coniferous forest of spruce in Idaho. (B) Temperate deciduous forest of oaks, hickories, and other hardwoods in Indiana. (Photographs courtesy of U.S. Forest Service.) (C) A tropical rain forest in Puerto Rico. (Photograph courtesy of University of Puerto Rico.)

species of these life forms are found in northern forests, but only in the tropical regions do they make up a conspicuous portion of the biological structure. Species diversity of both plants and animals tends to be high in tropical rain forests; there may be more species of plants and insects in a few acres of tropical rain forest than in the entire flora and fauna of Europe. Jordan (1971) points out that the ratio of leaf to new wood production is about 1:1 in tropical rain forests, compared with 1:6 in the temperate zone, which means tropical trees put proportionally more of their net production into leaves than into wood. Accordingly, annual leaf fall is greater in the tropics, but the energy content of leaves is less per unit of dry weight.

The unique characteristics of mineral cycling in tropical rain forests and their impact on the agricultural conversion of these forests were discussed in detail in Chapter 5. Rain forests are able to prosper on poor soils because of the many mutualistic and recycling adaptations within the biomass of the forest community. When the forest is removed, all these adaptive mechanisms are destroyed, and the land makes very poor cropland or pasture (Figure 15).

Two forest types that might be thought of as extremes in a moisture gradient are shown in Figure 16. Chaparral (Figure 16A) occurs in regions with winter rain and summer drought, and is a "fire type" in that it is naturally subjected to fires and is adapted to this factor (see Chapter 2). This kind of dwarf woodland is known as "macchie scrub" in the Mediterranean region and "mallee scrub" in Australia. Other types of dry-climate dwarf forests include the piñon-juniper of lower altitudes in the western mountains of the United States and the tropical thorn scrub forests of Africa. In contrast, temperate rain forests, such as those along the coast from northern California to Washington State (Figure 16B) occur where there is abundant moisture. They do not have as great a species diversity as tropical rain forests, but individual trees are larger, and the total timber volume may be greater. The California redwoods are a variant of this forest type.

A good place in which to observe the pattern of forests in relation to climate and substrata is the Great Smoky Mountains National Park, located along the Tennessee–North Carolina border. At sea level one would have to travel over hundreds of miles to observe the variety of climates present in a small geographic area in the Smokies. Figure 17 is a diagram that will help us to view the landscape with the eyes of the ecologist. The altitude change produces a north–south temperature gradient, whereas the valley and ridge topography provides a gradient of soil moisture conditions at any given altitude. The pattern of vegetation along the gradients stands out best in May and early June (when floral displays are also spectacular), but the remarkable way in which forests adapt to topography and climate is evident at any time of year.

(A)

(B)

FIGURE 15
Tropical rain forest. (A) A virgin stand. (B) When the rain forest is cleared, the land often makes poor pasture or cropland because the fertility is not in the soil but in the biomass that has been removed. (Photographs courtesy of Carl Jordan.)

(A)

(B)

FIGURE 16
Two forests adjusted to different moisture conditions. (A) Chaparral woodland, a dwarf forest of the winter rain–summer drought climate of coastal California; periodic fires are a major environmental factor. (B) A fine example of moist coniferous forest, often called temperate rain forest, in the Olympic National Forest of Washington State. Note the size of the large trees, the ground cover of ferns, and the epiphytic mosses that festoon the limbs of the trees. It is in this type of forest that the "clear-cut" controversies are raging. (Photographs courtesy of U.S. Forest Service.)

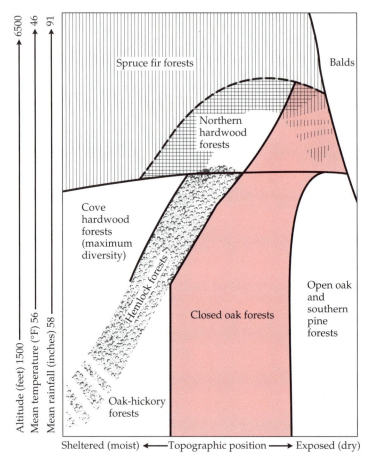

FIGURE 17

The pattern of forest vegetation in the Great Smoky Mountains National Park as related to temperature and moisture gradients. (Diagram prepared by R. Shanks after Whittaker 1952.)

As shown in Figure 17, the forests of the Smokies range from open oak and southern pine stands on the drier, warmer slopes at low altitudes, to northern coniferous forests of spruce and fir on the cold, moist summits. The southern pine stands extend upward along the exposed ridges, and the northern hemlock forest extends downward in the protected ravines where moisture and local temperature conditions are like those of higher altitudes. Maximum diversity of tree species occurs in sheltered (that is, moist) locations about midway in the temperature gradient.

The reason why some of the high, exposed slopes of the Smokies are covered with rhododendron thickets or grass instead of trees has not been adequately explained. These "balds" are not alpine tundras, for the altitude is not great enough for a true treeless zone. Whatever the reason (perhaps fire) for its original establishment, the shrub community is now so well established that it resists invasion by the forest. In this situation, we can observe how whole communities, as well as the individuals in them, compete with one another. The eventual outcome may depend on the occurrence of events such as fires or storms that might tip the balance in favor of one or the other ecological system.

Harvesting the Forests

Timber production and the practice of forestry pass through two phases. The first phase involves the harvest of net production that has been stored as wood over a period of years. When the accumulated growth of the past has been used up, we must adjust our forestry practice to harvest no more than the annual growth if we expect to have any wood products at all. In the northwestern United States, the first phase is still under way; the annual timber harvest in this region is about double the annual growth. In contrast, the second phase has been reached in the southeastern United States. Most of the old timber has been cut; hence, forestry practice is primarily concerned with young forests where the harvest now balances annual growth. Although annual net production in a young forest is often greater than that of an old forest, the quality of the wood for lumber use is not as good, since the wood of fast-growing young trees is not as dense as that of slow-growing older trees. As in so many situations, the dichotomy between quantity and quality has to be recognized; rarely can we have both. How timber is cut may depend on how timber is taxed. If standing timber is taxed on its market value every year, the owner (an individual or a timber company) is encouraged to harvest early in order to reduce taxes. If, on the other hand, the timber is not taxed until it is cut (and revenue obtained), there is an incentive to let the stand grow higher-quality wood.

In recent years there has been a raging controversy over massive clear-cuts of old-growth forests in the Pacific Northwest, especially in the national forests. Jerry Franklin of the University of Washington is a spokesman for a new generation of foresters who advocate managing a forest not just as a wood production system, but as an ecosystem with multiple uses and values, including watershed protection and life support. Reducing soil disturbance, leaving logs and debris in the forest, leaving uncut buffers along streams, leaving seed trees in stands being harvested, and arranging clear-cuts so as to provide corridors for animal use are some of the procedures being suggested. All of these ideas provide alternatives to the "stark choice between tree farms and total preservation" (Franklin 1989).

Caves

Very interesting heterotrophic communities are found in caves, which are numerous in many parts of the world. Organisms, many of them endemic—that is, found nowhere else—live for the most part on organic matter washed into the cave. Recently, Sarbu et al. (1996) explored a cave where a geothermal chemoautotrophic ecosystem had evolved. This system is similar to the undersea vent systems shown in Figure 2 in that hydrogen sulfide provides the energy source.

Our Forest-Edge Habitat

Human civilization seems to reach its most intense development in what was originally forest and grassland, especially in temperate regions. Consequently, most temperate forests and grasslands have been greatly modified from their primeval condition, but the basic nature of these ecosystems has by no means changed. Humans, in fact, tend to combine features of both grasslands and forests into habitats for ourselves that might be called **forest edge**. When humans settle in grassland regions, we plant trees around our homes, towns, and farms, so that small patches of forest become dispersed in what may have been treeless country. Likewise, when we settle in the forest, we replace most of it with grasslands and croplands (since little human food can be obtained from a forest), but leave patches of the original forest on farms and around residential areas. Many of the smaller plants and animals originally found in both forest and grassland are able to adapt and thrive in close association with humans and our domestic or cultivated species. The American robin, for example, once a bird of the forest, has become so well adapted to the human-made forest edge that it has not only increased in numbers, but has extended its geographic range. Most forest birds in Europe have switched from the forest to gardens, cities, and hedgerows—or else they have become extinct, since there are no longer many tracts of unbroken forest. Most native species that persist in regions heavily settled by humans become useful members of the forest-edge ecosystem, but a few become pests. The worst pests, however, are more likely to be species introduced from afar, as was discussed in Chapter 2.

If we consider croplands and pastures as modified grasslands of early successional types, then we can say that we depend on grasslands for food, but like to live and play in the shelter of the forest, from which we also garner useful wood products. At the risk of oversimplifying the situation, we might say that humans, in common with other heterotrophs, seek two basic things from the landscape: production and protection. But unlike lower organisms, we also find aesthetic enjoyment in the beauty of natural landscapes. For humankind, forests provide all three needs, but especially the latter two. In many cases the monetary value of wood, if harvested all at once, is far less than the value of intact forest that provides recreation, watershed protection, and other life-support services, home sites, and a modest harvest of wood as well (see Bergstrom and Cordell 1991).

Agroecosystems

Agroecosystems are domesticated ecosystems that are in many ways intermediate between natural ecosystems, such as grasslands and forests, and fabricated ecosystems, such as cities. They are solar-powered, as are natural ecosystems, but differ from them in several ways: the auxiliary energy sources that enhance productivity are processed fuels (along with animal and human labor) rather than natural energies; diversity is greatly reduced by human management in order to maximize the yield of specific foods or other products; the dominant plants and animals are under artificial rather than natural selection; and control is external and goal-oriented, rather than internal via subsystem feedback as in natural ecosystems (see Figure 11 in Chapter 3).

As noted in Chapter 1, plantation forests, like croplands, are agroecosystems—"tree farms" designed to increase wood and fiber production per unit of area (see Figure 2B in Chapter 1). Sooner or later the same management problems that affect other crops—loss of soil quality, pest control, artificial fertilizers to replace nutrients removed with the harvest, and so on—have to be dealt with. There is increasing interest in **agroforestry**, a practice that involves cultivation of small, fast-growing trees and food crops in alternate rows (MacDicken and Vergara 1990).

Agroecosystems resemble urban–industrial systems in their extensive dependence on and impact on externals; that is, both have large input and output environments. Agroecosystems differ from cities in being autotrophic rather than heterotrophic. The power density level (rate of energy flow per unit of area) of traditional agriculture, as practiced in economically undeveloped countries, is not much different from that of natural ecosystems. The power density of industrialized agriculture, however, is about ten times greater than that of most natural ecosystems due to the high energy and chemical subsidies needed. Its impact through agricultural chemical pollutants and soil erosion on waterways, the atmosphere, and other global life-support systems can approach in severity that of urban–industrial areas.

The properties of agroecosystems and the nature of their impact on other ecosystems have changed dramatically during the past half-century in the United States and other industrialized nations. It is important that we review this history in order to gain perspective on current problems and research needs. Auclair (1976) described the development of intensive agriculture in the midwestern United States in three stages as follows:

> From 1833 to 1934, some 90 percent of prairie, 75 percent of wetlands, and all forest land on good soils was converted to croplands, pastures, and wood lots. Natural vegetation was restricted to steep land and shallow, infertile soils. However, farms were generally small, crops diversified, and human and animal labor extensive, so the impact of farming on water, soil, and air quality was not deleterious overall.

> From 1934 to 1961, intensification of farming associated with inexpensive fuel and chemical subsidies, mechanization, and increase in crop specialization and monoculture occurred. Total cropland acreage decreased and forest cover increased 10 percent as more food was harvested from fewer acres by fewer farmers.

> From 1961 to 1980, energy subsidy, the size of farms, and farming intensity on the best soils all increased, with emphasis on continuous culture of grain and soybean cash crops, much of which is grown for export trade. Conservation practices, such as crop rotation, fallowing, terracing, and vegetated runoff waterways, decreased as farmers were forced to expand cash crops more and more to pay for the increasing costs of energy and machinery. Yields per unit were increased, but peaked for some grain crops during this period. Losses of farmlands to urbanization and soil erosion accelerated, as did the decline in water quality due to excessive fertilizer and pesticide runoff.

Brugam's (1978) analysis of the chemical composition of dated sections of cores from the bottom of a lake in Connecticut (Linsley Pond) provides a history of the changing impact of both agriculture and urbanization on adjacent ecosystems. Early farming in the 1800s had little effect on the lake, but intensification of agriculture after about 1915 caused eutrophication of the lake, resulting from an inflow of agricultural chemicals. From 1960 to the present, rapid urbanization and increased farming intensity has resulted in "hypereutrophication" due to agroindustrial wastes and extensive erosion that brings large amounts of soil, heavy metals, and other toxic substances into the lake. Marked changes in the biota directly related to changes in the input environment have been documented in this lake.

Given the increased cost of both energy and pollution, major technological, economic, and political efforts must be made to reduce the input costs of both agricultural and urban systems; otherwise, excesses in either or both will soon jeopardize the capacity of the natural life-support systems to support them. Viewing croplands, pasturelands, and plantation forestlands as dependent ecosystems that are functional parts of larger regional and global ecosystems (i.e., a hierarchical approach) is the first step in bringing together the disciplines necessary to accomplish these long-term goals. The so-called world food problem cannot be mitigated by the efforts of any one discipline, such as agronomy, working alone. Nor does ecology as a discipline offer any immediate or direct solutions, but the holistic and systems-level approaches that underlie ecological theory can make a contribution to the integration of disciplines.

Traditional versus Industrial Agroecosystems

Many long-standing traditional agricultural practices in less developed countries are receiving increasing attention because they are energy-efficient, ecologically sustainable, and provide adequate food for local people. These practices, however, do not produce excess grain or other products that can be exported as "commodities" for hard cash. Therefore, many small countries are pushing to replace their traditional agroecosystems with industrial ones, with little consideration of their downsides—namely, pollution and driving small farmers off the land and into the overcrowded cities.

Thailand is a good example of such a country. In rice culture agroecosystems, flood energy input is being replaced by industrial inputs of fossil fuel, fertilizers, and pesticides, resulting in increased yields per unit of area, but a reduction in energy efficiency (ratio of energy input to grain output) (Gajaseni 1995). Another example is Bali, where 1000-year-old gravity irrigation systems are being replaced by fuel-powered pumping systems, along with increased use of synthetic chemicals, for the sake of a short-term gain in market economics (Steven 1994).

In summary, market and other economic and political forces, along with urbanization and the pressures of human population growth, have transformed agroecosystems from "domesticated" ecosystems that were relatively harmonious with our general environment into increasingly "fabricated" ecosystems that more and more resemble urban-industrial ecosystems in energy and material demands and waste production. Fortunately, these trends are being reversed by what news magazine writer Hugh Sidey (1992) calls "revolution on the farm; the plow is being displaced by new technologies of 'residue management' that protect the land and promise even more abundant crops" (see Figure 15 in Chapter 5).

Urban–Industrial Techno-Ecosystems

Cities, suburbs, and industrial developments—that is, metropolitan districts, or what Neveh (1982) has called "techno-ecosystems"—are small but very energetic islands in the matrix of natural and agricultural landscapes (Dornet and McClellan 1984). As emphasized in Chapter 1, these fabricated environments are parasites on the biosphere; that is not meant to belittle cities, but to be realistic.

Cities are supposed to be the crowning achievements of humankind. The planned cites of ancient Greece, the city-states of the Middle Ages, and the towering skyscrapers of New York and other large cities today all inspire wonder and awe. Future cities such as those depicted at Disney World are technological utopias where everyone is rich and lives in complete comfort with lots of time for play. Unfortunately, most cities today are in disorder and decline, even as people all over the world rush to the cities seeking a better economic life. The growth of cities is especially rapid in the less developed countries. By the year 2000 Mexico City and São Paulo, Brazil, are projected to have more than 25 million people each—far more than any city in the industrialized world except possibly Tokyo, and approaching twice as many as New York. It is also predicted that by the year 2010, between 50 and 80 percent of people worldwide will be urban (see the 1994 United Nations report edited by Fuchs et al.).

As documented in Chapter 6, anything that grows rapidly and haphazardly (without plan or control) and without regard to life support will outstrip the infrastructure necessary to maintain its growth, thereby bringing on boom-and-bust cycles. Whether we as free-spirited individuals like it or not, we as a society must engage in some serious urban planning. Most of the literature dealing with the plight of the cities focuses on internal problems such as deteriorating infrastructure and crime, but as John T. Lyle (1985, 1993) points out, cities of the future will have to "embrace the ecology of the landscape, rather than set themselves apart." Urban regeneration, then, will depend more and more on reconnecting the city to the life-supporting land, because as we pointed out early in this book, a parasite prospers only if its host remains in good condition (see also Haughton and Hunter 1994). What we can learn from ecology about dealing with this situation will be discussed in the Epilogue.

Suggested Readings

BIOGEOGRAPHY

Brown, J. H., and A. C. Gibson. 1983. *Biogeography*. Mosby, St. Louis.

Cox, C. B., I. N. Healey, and P. D. Moore. 1973. *Biogeography: An Ecological and Evolutionary Approach*. 2nd ed. Blackwell Scientific, Oxford.

Hallam, A. 1972. Continental drift and the fossil record. *Sci. Am.* 227(5): 56–66.

MacArthur, R. H. 1972. *Geographical Ecology*. Harper & Row, New York.

Pielou, E. C. 1979. *Biogeography*. John Wiley, New York.

THE OCEAN

Barber, R. T., and R. L. Smith. 1980. Coastal upwelling ecosystems. In *Analysis of Marine Ecosystems*, ed. A. R. Longhurst. Academic Press, New York.

Bretherton, F. P., ed. 1986. Changing climates and the oceans. *Oceanus* 29(4) (special issue). (Twelve articles, several illustrated in color.)

Carson, R. 1952. *The Sea Around Us*. Oxford University Press, New York.

*Cloud, P. E. 1969. *Resources and Man*. Freeman, San Francisco.

Falkowski, P. G., ed. 1980. *Primary Productivity in the Sea*. Plenum Press, New York. (See also the 1981 review in *Science* 212: 794.)

*Heezen, B. C., C. M. Tarp, and M. Ewing. 1959. *The Floors of the Ocean. I. North Atlantic*. Special Paper no. 65., Geological Society of America, Boulder, CO.

MacIntyre, F. 1970. Why the sea is salt. *Sci. Am.* 223(5): 104–115.

Odum, E. P. 1971. *Fundamentals of Ecology*, 3rd ed. Chapter 12. Saunders, Philadelphia.

*Pomeroy, L. R. 1974. The ocean's food web, a changing paradigm. *BioScience* 24: 499–504.

*Pomeroy, L. R. 1984. Significance of microorganisms in carbon and energy flow in aquatic ecosystems. In *Current Perspectives in Microbial Ecology*, eds. M. J. Lug and C. A. Reddy, pp. 405–411. American Society of Microbiologists, Washington, D.C.

Revelle, R., ed. 1969. The ocean. *Sci. Am.* 221(3) (special issue).

*Smith, C. R., H. Kukert, R. A. Weatcroft, P. A. Jumers, and J. W. Deming. 1989. Vent fauna on whale remains. *Nature* 341: 27–28.

*Swan, L. W. 1992. The Aeolian biome: Ecosystems of the earth's extremes. *BioScience*, 42: 262–270.

Thorson, G. 1971. *Life in the Sea*. McGraw-Hill, New York.

*Tunnicliffe, V. 1992. Hydrothermal vent communities of the deep sea. *Am. Sci.* 80: 336–349. (See also back-to-back articles by Grassle and by Jannasch and Mottl in *Science* 229: 713–725, 1985.)

COASTAL WATERS, ESTUARIES, AND SEASHORES

Amos, W. H. 1966. *The Life of the Seashore*. Our Living World of Nature series. McGraw-Hill, New York.

Carson, R. 1956. *The Edge of the Sea*. Houghton Mifflin, Boston.

Goldreich, P. 1972. Tides and the earth–moon system. *Sci. Am.* 226(4): 42–52.

Kaufman, W., and O. H. Pilkey, Jr. 1983. *The Beaches Are Moving*. Duke University Press, Durham, NC.

MacLeish, W. H., ed. 1976. Estuaries. *Oceanus* 19(5) (special issue). (Articles by 10 authors explore various aspects of estuaries.)

Mann, K. H. 1982. *Ecology of Coastal Waters*. Studies in Ecology, vol. 8. University of California Press, Berkeley.

*Muscatine, L., and J. W. Porter. 1977. Reef corals: Mutualistic symbioses adapted to nutrient-poor environments. *BioScience* 27: 454–460.

Odum, E. P. 1961. The role of tidal marshes in estuarine production. *The Conservationist*, June–July, 12–15.

Odum, E. P. 1980. The status of three ecosystem-level hypotheses regarding salt marsh estuaries. In *Estuarine Perspectives*, ed. V. S. Kennedy. Academic Press, New York.

Odum, W. E. 1970. Insidious alteration of the estuarine environment. *Trans. Am. Fish. Soc.* 90: 836–847.

*Odum, W. E., and C. C. McIvor. 1990. Mangroves. In *Ecosystems of Florida*, eds. R. J. Myers and J. J. Ewel, pp. 517–548. University of Central Florida Press, Orlando.

Pearse, A. S., H. J. Humm, and G. W. Wharton. 1942. Ecology of sand beaches. *Ecol. Monogr.* 12: 136–190.

Stephenson, T. A., and A. Stephenson. 1973. *Life Between Tidemarks on Rocky Shores*. W. H. Freeman, San Francisco.

Teal, J., and M. Teal. 1969. *Life and Death of the Salt Marsh*. Little, Brown, Boston.

Warner, W. W. 1976. *Beautiful Swimmers: Watermen, Crabs, and the Chesapeake Bay*. Penguin Books, New York.

Wiegert, R. G., and B. J. Freeman. 1990. *Tidal Salt Marshes of the Southeastern Atlantic Coast: A Community Profile*. Biological Report no. 86(7.29), Fisheries & Wildlife Service, Washington, D.C. (See also W. E. Odum, 1988 in the next section.)

FRESH WATERS AND WETLANDS

Baxter, R. M. 1977. Environmental effects of dams and impoundments. *Annu. Rev. Ecol. Syst.* 8: 255–284.

*Belt, C. B. 1975. The 1973 flood and man's constriction of the Mississippi River. *Science* 189: 681–684.

Cummins, K. W. 1974. Structure and function of stream ecosystems. *BioScience* 24: 631–641.

Cushing, C. E., and K. W. Cummins, eds. 1995. *Ecosystems of the World*. Vol. 22. *River and Stream Ecosystems*. Elsevier, Amsterdam.

*Davis, S. M., J. C. Ogden, and W. A. Parks, eds. 1995. *Everglades: The Ecosystem and Its Restoration*. St. Lucie Press, Delray Beach, FL. (Papers by 57 authors.)

Deevey, E. S. 1951. Life in the depths of a pond. *Sci. Am.* 185(4): 68–72.

*Edmondson, W. T. 1968. Water quality management and lake eutrophication: The Lake Washington case. In *Water Resources Management and Public Policy*. University of Washington Press, Seattle. (A pollution abatement success story; see also Sunday supplement, *Seattle Times*, Aug. 4, 1985, and Edmondson 1970.)

Edmondson, W. T. 1970. Phosphorus, nitrogen, and algae in Lake Washington after diversion of sewage. *Science* 169: 690–691.

Eliassen, R. 1952. Stream pollution. *Sci. Am.* 186(3): 17–21.

Fisher, S. G., and G. E. Likens. 1972. Stream ecosystem: Organic energy budget. *BioScience* 22: 33–35.

*Gasaway, C. R. 1970. *Changes in the Fish Population of Lake Francis Case in South Dakota in the First 16 Years of Impoundment.* Technical Paper no. 56. Bureau of Sport Fisheries and Wildlife, Washington, D.C.

Good, R. E., D. F. Whigham, and R. L. Simpson. 1978. *Freshwater Wetlands: Ecological Processes and Management Potential.* Academic Press, New York.

*Greeson, P. E., J. R. Clark, and J. E. Clark, eds. 1979. *Wetland Functions and Values: The State of Our Understanding.* American Water Resources Association, Minneapolis. (All you want to know about wetlands, and more!)

*Holeman, J. N. 1968. The sediment yield of major rivers of the world. *Water Res.* 4: 737–747.

Kitchell, J. F. 1992. *Food Web Management: A Case Study of Lake Mendota.* Springer-Verlag, New York. (Fish and algal blooms in one of the most intensively studied North American lakes.)

*Lodge, T. E. 1994. *The Everglades Handbook: Understanding the System.* St. Lucie Press, Delray Beach, FL.

Lugo, A. E., S. Brown, and M. M. Brinson. 1988. Forested wetlands in freshwater and saltwater environments. *Limnol. Oceanogr.* 33(4, part 2): 894–909.

Miller, J. A., ed. Ecology of large rivers. Special issue of *BioScience*, 45: 134–203.

*Mitsch, W. J., and J. G. Gosselink. 1994. *Wetlands*, 2nd ed. Van Nostrand-Reinhold, New York.

Niering, W. A. 1966. *The Life of the Marsh.* Our Living World of Nature series. McGraw-Hill, New York.

*Odum, E. P. 1988. Wetland values in retrospect. In *Freshwater Wetlands and Wildlife*, eds. R. R. Sharitz and J. W. Gibbons, pp. 1–8. Department of Energy Symposium Series no. 61. Office of Technological Information, Oak Ridge, TN. (See also E. P. Odum 1983, *J. Soil Water Conserv.* 38: 546–549.)

*Odum, W. E. 1988. Comparative ecology of tidal freshwater and salt mashes. *Annu. Rev. Ecol. Syst.* 19: 137–176.

Patrick, R. 1970. Benthic stream communities. *Am. Sci.* 58: 546–549.

Porter, K. G. 1977. The plant–animal interface in freshwater ecosystems. *Am. Sci.* 65: 159–170.

Ragotzkie, R. A. 1974. The Great Lakes rediscovered. *Am. Sci.* 62: 454–464.

Smith, R. A., R. B. Alexander, and M. G. Wolman. 1987. Water-quality trends in the nation's rivers. *Science* 235: 1607–1615.

Wolman, M. G. 1971. The nation's rivers. *Science* 174: 905–918. (Excellent graphs and tables on water quality indices.)

THE TERRESTRIAL BIOMES

Aber, J. D., and J. M. Melillo. 1991. *Terrestrial Ecosystems.* Saunders, Philadelphia.

Allen, D. L. 1967. *The Life of Prairies and Plains.* Our Living World of Nature series. McGraw-Hill, New York.

Barbour, M. G., and W. D. Billings. 1988. *North American Terrestrial Vegetation.* Cambridge University Press, New York.

*Bergstrom, J. C., and H. K. Cordell. 1991. An analysis of the demand for and value of outdoor recreation in the United States. *J. Leisure Res.* 23: 67–86. (The economic value of the National Forests to wilderness and recreation users— $122 billion per year—is nine times the aggregate value of round wood timber harvest.)

Bormann, F. H., and G. E. Likens. 1979. *Pattern and Process in a Forested Ecosystem.* Springer-Verlag, New York. (Synthesis of research at Hubbard Brook in New Hampshire, a pioneer ecosystem-level watershed study.)

Caufield, C. 1984. *In the Rainforest: Report from a Strange, Beautiful, Imperiled World.* University of Chicago Press, Chicago.

Cox, T. R., R. S. Maxwell, P. D. Thomas, and J. J. Malone. 1985. *This Well-Wooded Land: Americans and Their Forests from Colonial Times to the Present.* University of Nebraska Press, Lincoln.

Denison, W. C. 1973. Life in tall trees. *Sci. Am.* 228(6): 74–80.

Douglas, I. 1967. Man, vegetation, and sediment yields of rivers. *Nature* 215: 925–928.

*Franklin, J. 1989. Towards a new forestry. *American Forests* November–December, pp. 37–44. (There is an alternative to the stark choice between tree farms and total preservation.)

Franklin, J. F. 1993. Lessons from old growth. *J. Forestry* 91: 10–13.

Hadley, N. F. 1972. Desert species and adaptation. *Am. Sci.* 60: 338–347.

Hunt, C. B. 1973. *Natural Regions of the United States and Canada.* W. H. Freeman, San Francisco.

*Jordan, C. 1971. A world pattern in plant energetics. *Am. Sci.* 59: 426–433.

McCormick, J. 1966. *The Life of the Forest.* Our Living World of Nature series. McGraw-Hill, New York.

McNaughton, S. J., M. Oesterheld, D. A. Frank, and K. J. Williams. 1991. Primary and secondary production in terrestrial ecosystems. In *Comparative Analysis of Ecosystems,* eds. J. Cole, G. Lovett, and S. Findley. Springer-Verlag, New York. (Primary production is strongly influenced and sometimes controlled by herbivores and other secondary producers, and is an integrated variable of processes throughout the ecosystem.)

*Morello, J. 1970. Modelo de relaciones entra pastizales y lenosas colonzodores en el Chaco Argentino. (A model of relationships between grassland and woody colonizer plants in the Argentine Chaco.) *Idia* 276: 31–51.

Richards, P. W. 1973. The tropical rain forest. *Sci. Am.* 229(6): 58–67.

*Sarbu, S. M., T. C. Kane, and B. K. Kinkle. 1996. A chemoautotrophically based cave ecosystem. *Science* 272:1953–1955.

*Schmidt-Nielsen, K. 1973. *Desert Animals: Physiological Problems of Heat and Water.* Oxford University Press, Oxford.

Shelford, V. E. 1963. *The Ecology of North America.* University of Illinois Press, Urbana.

Shelford, V. E., and S. Olson. 1935. Sere, climax, and influent animals with special reference to the transcontinental coniferous forest of North America. *Ecology* 16: 375–402. (A classic paper documenting how animals link together various developmental stages of vegetation within a major biome type.)

Shugart, H. H., and D. C. West. 1981. Long-term dynamics of forest ecosystems. *Am. Sci.* 69: 647–652.

Sinclair, A. R. E., and M. Norton-Griffiths, eds. 1979. *Serengeti: Dynamics of an Ecosystem.* University of Chicago Press, Chicago.

Sutton, A., and M. Sutton. 1966. *The Life of the Desert.* Our Living World Of Nature series. McGraw-Hill, New York.

Terborgh, J. 1992. *Diversity and the Tropical Rain Forest.* Scientific American Library, W. H. Freeman, New York. (Beautifully illustrated, moderately technical account of the legendary tropical rain forest.)

Walter, H. 1973. *Vegetation of the Earth and Ecological Systems of the Geo-biosphere,* 3rd ed. Springer-Verlag, New York.

Waring, R. H., and J. F. Franklin. 1979. Evergreen coniferous forests of the Pacific Northwest. *Science* 204: 1380–1386.

Waring, R. H., and W. H. Schlesinger. 1985. *Forest Ecosystems: Concepts and Management.* Academic Press, Orlando, FL.

*Whittaker, R. H. 1952. Vegetation of the Great Smoky Mountains. *Ecol. Monogr.* 26: 1–80.

AGROECOSYSTEMS

Altieri, M. A. 1987. *Agroecology: The Scientific Basis of Alternative Agriculture.* Westview Press, Boulder, CO. (Contrasts chemical–mechanized agriculture with traditional farming in underdeveloped countries and conservation tillage emerging in developed countries.)

*Auclair, A. N. 1976. Ecological factors in the development of intensive-management ecosystems in the midwestern United States. *Ecology* 57: 431-444.

*Brugam, R. B. 1978. Human disturbance and the historical development of Linsley Pond. *Ecology* 59: 19-36.

Carroll, C. R., J. H. Vandermeer, and P. M. Rosset, eds. 1990. *Agroecology.* McGraw-Hill, New York.

Dover, M. J., and L. M. Talbot. 1987. *To Feed the Earth: Agro-ecology for Sustainable Development.* World Resources Institute, Washington, D.C.

Edwards, C. L., R. Lal, P. Madden, R. H. Miller, and G. House, eds. 1990. *Sustainable Agricultural Systems.* St. Lucie Press, Delray Beach, FL.

*Gajaseni, J. 1995. Energy analysis of wetland rice systems in Thailand. *Agr. Ecosyst. Environ.* 52: 173–178.

Gliessman, S. R., ed. 1990. *Agroecology: Researching the Ecological Basis for Sustainable Agroecology.* Springer-Verlag, New York.

Lowrance, R., B. R. Stinner, and G.J. House, eds. 1984. *Agricultural Ecosystems: Unifying Principles.* John Wiley, New York.

*MacDicken, K. G., and N. T. Vergara. 1990. *Agroforestry.* John Wiley, New York.

Sidey, H. 1992. Revolution on the farm. *Time* 139(26): 54–56.

*Steven, J. E. 1994. Science and religion at work. *BioScience* 44: 60–64. (Replacement of long-term ecologically sound irrigation system with commodity-based market system.)

Worster, D. 1990. Transformation of the earth: Towards an agroecological perspective in history. *Am. Hist.* 76: 1087–1106.

URBAN–INDUSTRIAL TECHNO-ECOSYSTEMS

Dornet, R. S., and P. W. McClellan. 1984. The urban ecosystem, its spatial structure, its subsystem attributes. *Environment* 16(1): 9–20.

*Fuchs, R. L., E. Prennon, J. Chamie, F. C. Lo, and J. L. Vito. *Mega-City Growth and the Future.* United Nations Publishing, New York.

*Haughton, G., and C. Hunter. 1994. *Sustainable Cities.* Jessica Kingsley, London.

Lowe, M. D. 1991. *Shaping Cities: The Environmental and Human Dimensions.* Worldwatch Paper no. 105. Worldwatch Institute, Washington, D.C.

*Lyle, J. T. 1985. *Design for Human Ecosystems.* Van Nostrand Reinhold, New York.

*Lyle, J. T. 1993. Urban ecosystems. *In Context* 35: 43–45.

Morris, D. 1982. *Energy and the Transformation of Urban America.* Sierra Club Books, San Francisco.

Neveh, Z. 1982. Landscape ecology as an emerging branch of human ecosystem science. *Adv. Ecol. Res.* 12: 189–237.

Vining, D. R. 1985. The growth of core regions in the third world. *Sci. Am.* 253(4): 42–49. (Explosive growth of third world cities is creating intractable economic, environmental, and social problems.)

*Indicates references cited in this chapter.

Epilogue

The Transition from Youth to Maturity

> Economic deficits may dominate our headlines, but ecological
> deficits will dominate our future.
> —Lester Brown et al., *The State of the World*, 1986

Predicting the future is a fascinating game that is especially popular in
times of crisis. Actually, no one can predict the future with any degree of
precision—there are just too many unknowns, too many new events, technological innovations, and other factors that cannot be foreseen. Nevertheless, it is instructive to consider a range of possibilities that could come
to pass. We then may be able to estimate their probability, given current
conditions, understanding, and knowledge. Most important, we might be
able to do something now to reduce the probability of an undesirable
future.

As we approach the year 2000, about the only certainties are that:
human beings will continue to increase in numbers, at least until well into
the next century (see Figure 5 in Chapter 6); something will have to be
done about the fouling of our life-support systems (especially the atmosphere); humanity will have to make a major and very painful transition in
energy use from fossil fuels to other less certain, and probably less lucrative, sources; and finally, humanity will overshoot its safe carrying capacity, as we seem to be already doing in many ways and with many

resources, bringing on boom-and-bust cycles. The challenge of the future, then, will be not how to avoid the overshoot, but how to survive it.

Most futurists believe that we need to reduce our current prodigious waste and become more efficient in order to do more with less high-quality energy and reduce the pollution that results from energy waste. Most also agree that reducing per capita energy consumption in the industrialized countries would not only *improve* their quality of life (Nader and Beckerman 1978), but also help the poorer nations improve theirs.

Most students of the future also agree that rapid growth should be avoided if for no other reason than that it tends to create social and environmental problems faster than they can be dealt with. Rapid population growth and urban–industrial development combine to create a momentum that is very difficult to control (see National Academy of Sciences 1971 and Catton 1980). It is significant that in 1992 the prestigious U.S. National Academy of Sciences and the Royal Society of London issued the following joint proclamation:

> World population is growing at the unprecedented rate of almost 100 million people every year, and human activities are producing major changes in the global environment. If current predictions of population growth prove accurate and patterns of human activity remain unchanged, science and technology may not be able to prevent either irreversible degradation of the environment or continued poverty for much of the world.

There is no shortage of studies, "think tank" reports, or popular books assessing the current predicament of humankind. Many paint a rather grim picture of present global problems, but others are optimistic about the future. The ways in which scholars, as well as people in general, view the future range from complete confidence in business as usual and new technology (a "more of the same" philosophy) to a belief that society must completely reorganize, "power down," and develop new international and holistic political and economic procedures in order to deal with a world of finite resources. The late Herman Kahn (*The Next 200 Years*, 1976) and economist Julian Simon (*The Ultimate Resource*, 1981) are well-known spokesmen for the former view, while biologist Paul Ehrlich (*The Population Bomb*, 1968), E. F. Schumacher (*Small Is Beautiful*, 1973), Kenneth Watt (*The Unsteady State*, 1977), physicist Fritjof Capra (*The Turning Point*, 1982), and economist Herman Daly (*Toward a Steady-State Economy*, 1973) are among those arguing for the need for fundamental changes, a position that is becoming more and more of a consensus among world leaders. Then there are the cornucopian ("horn-of-plenty") technologists who are optimistic that an efficient and clean hydrogen economy (to replace carbon-dirty fossil fuels), landless agriculture, wasteless industry, and other future technologies will enable 10 billion or more people to coexist with enough natural environment to provide for the necessary life-support, preservation of endangered species, and enjoyment of nature (see Ausubel 1996).

Global Models

Some of the most comprehensive futuristic reports, the work of many collaborators, include those prepared by the Club of Rome and the global models produced by the United States and other governments and the United Nations. The Club of Rome is a group of scientists, economists, educators, humanists, industrialists, and civil servants brought together by Italian industrialist Dr. Arillio Peccei, who felt the urgent need to prepare a series of books on the future predicament of humankind. Its first and best-known book, *The Limits to Growth* (Meadows et al. 1972), predicted on the basis of models that if our present political and economic methods continue unchanged, severe boom-and-bust cycles will occur. Essentially, this first Club of Rome study employed a modern systems approach to the older "warnings-to-humankind" classics such as George Perkins Marsh's *Man and Nature* (1864, reissued in 1965), Paul Sears's *Deserts on the March* (1935), William Vogt's *Road to Survival* (1948), Fairfield Osborn's *Our Plundered Planet* (1948), and Rachel Carson's *Silent Spring* (1962). The report

---

Buzzwords: Sustainability versus Maturity

The words "sustainable" and "sustainability," used to describe our goal for society, are appearing with increasing frequency in articles, editorials, and books, but with quite a wide range of meanings. "Sustainable growth" or "sustainable development," for example, can be read as maintaining balances and resources into the future, or as continuing to sustain growth forever (i.e., bigger is always better), an obvious impossibility. I prefer the word "maturity" to describe our goal for society because we all experience the difficult transition from physical growth to maturity, and thereby understand what is really involved in going from quantitative to qualitative development (i.e., from getting bigger to getting better)—hence the subtitle of this Epilogue.

denounced society's obsession with growth, in which at every level (individual, family, corporation, and nation) the goal is to get richer and bigger and more powerful, without considering basic human values and the ultimate cost of unrestricted, unplanned consumption of resources and stress on environmental life-support goods and services.

The Limits to Growth was followed by a series of additional reports that attempted not only to describe present situations and possible future trends in more detail but also to suggest actions that should be taken to avoid a boom-and-bust doomsday. These studies were published as books with titles such as *Mankind at the Turning Point, Reshaping the International Order, Goals for Mankind, Wealth and Welfare,* and *No Limits to Learning: Bridging the Human Gap* (all published by Pergamon Press, New York). A variety of distinguished scholars contributed to these efforts, including engineers, economists, philosophers, historians, and educators. Laszlo (1977) assesses the overall impact of the reports as follows:

> Thanks largely to the efforts of the Club of Rome, international awareness of the world problematique has rapidly grown. The Club pioneered the way (to use a medical analogy) from diagnosis to prescription but very little was accomplished in the way of therapy. To use another metaphor, the Club helped point the way but did little to generate the will to take it.

Between 1971 and 1981, a number of other global models were developed. These models are computerized mathematical simulations of the world's physical and socioeconomic systems, and make projections into the future that are logical consequences of the data and the assumptions that went into the model. It should be emphasized that each model is unique with respect to the assumptions that motivated it. These models have been reviewed and compared as a group in a report issued by the Congressional Office of Technology Assessment (OTA 1982), and in a book by Donella Meadows et al. (1982).

Despite differing assumptions and biases, the models as a set do agree on some points, namely:

1. Technological progress is expected, and is vital, but social, economic, and political changes will also be necessary.
2. Populations and resources cannot grow forever on a finite planet.
3. A sharp reduction in the growth rates of population and urban–industrial development will greatly reduce the probability of overshoots or major breakdowns in life-support systems.
4. Continuing "business as usual" will not lead to a desirable future, but rather will result in further widening of undesirable gaps (rich–poor, for example).

5. Cooperative long-range approaches will be more beneficial for all parties than competitive short-term policies.
6. Because the interdependence among peoples, nations, and the environment is much greater than commonly imagined, decisions should be made within a holistic (systems) context. Actions to alter current undesirable trends (atmospheric toxification, for example) taken soon (within the next couple of decades) will be more effective and less costly than actions taken later. This calls for strong political leadership and vigorous public education, since by the time a problem is obvious to everybody, it may be too late.

In the 1990s Donella Meadows and colleagues cranked up their computers again and came out with a sequel to *The Limits to Growth* entitled *Beyond the Limits* (1992). They conclude that global conditions are worse than predicted in 1972; however, they still envision a sustainable future if the six points outlined above are taken seriously and acted on. They comment that what the world needs is good old-fashioned "love" that will enable people to work together for common causes. In Chapter 7 we discussed a parallel trend in ecosystem development, in which mutualism increases when resources are scarce.

So much for models. They are good for integrating data and trends, but they are not able to factor in human resolve and ingenuity (or lack of it). The rest of the Epilogue will explore several approaches to understanding our predicament and options for getting out of the mess we are in.

Worrisome Gaps

A good way to assess the predicament of humankind is to consider the gaps that must be narrowed if humans and the environment, as well as nations, are to be brought into more harmonious relationships. Among these gaps, several of which have already been mentioned in Chapter 4 of this book, are the following:

1. The income gap: between the rich and the poor, both within nations and between industrialized nations (30 percent of the world's population) and the nonindustrialized nations (70 percent)
2. The food gap: between the well-fed and the underfed
3. The value gap: between market and nonmarket goods and services
4. The education gap: between the literate and the illiterate, the skilled and the unskilled

None of these gaps has been narrowed very much during the past several decades; in fact, the income and value gaps have gotten much worse. According to Seligson (1984), the gap in per capita income between rich and poor nations grew from $3,617.00 to $9,648.00 between 1950 and 1980; the gap continues to grow in the 1990s. Well-meaning efforts by wealthy nations to help poorer ones have too often failed because the deleterious cultural and environmental impacts of the aid were not anticipated. For example, construction of a reservoir in a fertile valley may provide benefits initially, but may also force farmers to move upstream to less suitable land, resulting in severe erosion and deforestation of the watershed and subsequent silting in the reservoir. Some 20 years ago Morehouse and Sigurdson (1977) pointed out that the transfer of industrial technology from industrialized to developing nations too often benefits the small modern sector but not the masses of rural poor in the traditional society. Wealth does not "trickle down" when there are profound cultural, educational, and resource differences within the population. As noted in Chapter 4, one cannot transfer a high-energy industrial or agricultural technology to a poor country without also providing the high-quality energy needed to sustain it. It may be better to enhance current "low-tech" operations until the nation can "go high-tech" on its own. As in nature, one gets to the climax stage only after the way is prepared by developmental stages.

The Ecological Assessment

The wisdom of the many contributors to the Club of Rome reports, as well as the output of global models, conforms rather well to basic ecosystem theory, especially three of its paradigms: a holistic approach is necessary when dealing with complex systems; cooperation has greater sur-

The Greatest Good . . .

Not too long ago, when the world was not so crowded, we heard a lot about "the greatest good for the greatest number" as the goal for the development of human society. But we are finding out by experience that the greatest good in terms of quality of life for the individual comes when the numbers are not the greatest! As we have pointed out in this book, ecological studies do provide a basis for placing society's "pursuit of happiness" goal on a qualitative rather than a quantitative basis (see, for example, the box "Errington's Muskrats" in Chapter 6.

vival value than competition when limits (resources or otherwise) are approached; orderly, high-quality development of human communities, like that of biotic communities, requires negative as well as positive feedback. As I have noted elsewhere (E. P. Odum 1977), these scholars' conclusions also conform to the age-old human wisdom found in common-sense proverbs, such as "look before you leap," "don't put all your eggs in one basket," "haste makes waste," "an ounce of prevention is worth a pound of cure," "power corrupts," and many more.

A civilization is a system, not an organism, contrary to what Arnold Toynbee said in *A Study of History* (1961). Civilizations do not necessarily have to grow, mature, become senescent, and die as organisms do, even though this process has happened in the past (the rise and fall of the Roman Empire, for example). According to geographer Karl Butzer (1980), civilizations become unstable and break down when the high cost of maintenance results in a bureaucracy that makes excessive demands on the productive sector. Such a view coincides with ecological theories of carrying capacity (Chapter 6), energy flow (Chapter 4), and complexity (Chapter 3). As we keep pointing out, we can learn a lot from the study of ecology that will help us deal with human predicaments.

Historical Perspectives

Anthropologist Brock Bernstein (1981) notes that in many isolated cultures that must survive on local resources alone, actions that would be detrimental to the future are perceived and avoided. Such local feedback in decision-making is lost when isolated cultures are incorporated into large and complex industrial societies. As Bernstein says, "Economics must develop a coherent theory of decision-making behavior that is applicable at all levels of group organization. This will necessitate defining self-interest in terms of survival rather than consumption." Such a shift would bring economic behavior under something akin to natural selection, which has worked so well to insure the perpetuation of life on earth over the eons.

One of the obstacles to avoiding overshoots in resource use is what Garrett Hardin calls **the tragedy of the commons** (Hardin 1968). Hardin is a professor of human ecology who thinks deeply and writes eloquently on our population and environmental dilemmas. By "commons" he means that part of our environment that is open to use by anyone and everyone, with no one person responsible for its welfare. A pasture or open range shared by many herdsmen is an example. Since it is to the advantage of each herdsman to graze as many cattle as possible, the capacity of the range to sustain grazing will be exceeded unless restrictions are agreed upon and enforced by the community as a whole. Many commons were protected by such community-enforced restrictions and customs prior to the Industrial Revolution. Primitive herding societies solved the problem by moving their cattle from one place to another on a regular basis before overgrazing occurred at any one place. Many European cities have a long tradition of maintaining commons in the form of large parks and green belts. The "tragedy" in these modern times is that local restrictions, as might be embodied in zoning ordinances, are so easily overturned by the pressure of "big money"—that is, the capital that is available for the kind of development that yields large short-term profits, often at the expense of local quality of life. In too many cities, citizens have to engage in constant (and too often losing) battles to keep their neighborhoods from being overbuilt.

In another book, Hardin (1985) raises a most intriguing question: Would the Industrial Revolution have gotten off the ground without exploitation of people and environment in the beginning? Recall Dickens's novels describing labor abuse and the complete inattention to air and water pollution in the nineteenth century. Certainly, exploitation of people (e.g., industrial sweatshops) and unrestricted pollution of the environment greatly accelerated the capital accumulation on which the present affluence of the industrial world is based. But while it may have been justifiable to exploit people and the environment in the early stages of development in order to build material wealth, we are beginning to realize (as Hardin is quick to point out) that we are at a turning point in history. We cannot continue "commonizing the costs and privatizing the profits" (Hardin's words) and postponing the environmental and human costs of rapid growth and development without incurring widespread damage to our global life-support systems. **Growth management** is a new "buzzword" that can be used to open communication among the disciplines and special interest groups that must be involved in developing the new political and economic infrastructures needed to protect the quality of life.

Social Traps

A situation in which a short-term gain is followed in the long term by a costly or deleterious situation not in the best interest of either the individual or society has been called a **social trap** (Platt 1973; Cross and Guyer 1980). An analogy is a trap that entices an animal into it with an attractive bait; the animal enters the trap in the hope of an easy meal, but then finds it difficult or impossible to escape. Cigarette smoking is an example of a behavioral social trap, while hazardous waste dumping, destruction of wetlands (or other life-support environments), and nuclear war are examples of environmental social traps. Edney and Harper (1978) suggest using a simple game played with poker chips to illustrate the relationship of social traps to the tragedy of the commons. A pool of poker chips is established, and each player has a choice of removing from one to three chips. The pool is renewed after each round in proportion to the number of chips remaining. If players think only of their immediate, short-term gains and remove the maximum of three chips, the renewable resource of the common pool becomes smaller, and ultimately the resource pool is gone. Removing one chip each round sustains the renewable resource. This game enables "kids of all ages" to experience the folly of being greedy.

Cross and Guyer (1980) and Costanza (1987) suggest turning social traps into **trade-offs** by levying a tax or charge on the parties responsible for creating long-term deleterious situations—for example, levying a pollution tax on a hazardous waste generator. Money collected in this manner could be put into a trust fund and used to monitor and ameliorate environmental impacts; if the impact was less deleterious than originally predicted, money could be returned to the generators or future taxes could be reduced.

Coming Full Circle

As we discussed in Chapter 7, human societies go from pioneer to mature status in a manner parallel to the way that natural communities undergo ecological succession and individuals go from youth to adulthood. As we have already noted, there are many strategies and behaviors that are appropriate and necessary for survival during youth but which become inappropriate and detrimental at maturity. Continuing to act on a short-term, one-problem/one-solution basis as society grows larger and more complex leads to what economist A. E. Kahn (1966) calls "the tyranny of small decisions." Increasing the heights of smokestacks—a quick fix for local smoke pollution—is an example in which many such "small decisions" lead to a larger problem of increased regional air pollution (see Figure 7 in Chapter 5). W. E. Odum (1982) gives another example: No one purposefully planned to destroy 50 percent of the wetlands along the northeastern coast of the United States between 1950 and 1970, but it happened, as a result of hundreds of small decisions to develop small tracts of marshland. Finally the state legislatures woke up to the fact that valuable life-support environment was being destroyed, and one by one they passed wetlands protection acts in an effort to save the remaining wetlands.

What all this means for the future is that the transition time for human communities has come, or will be coming soon, that necessitates "coming full circle" or "doing an about-face" on many previously acceptable concepts and procedures.

In the next eight sections we will present the growth–maturity theme from different but interrelated viewpoints, with suggestions for solutions to our dilemmas. We will conclude with the concept of dual capitalism and two global survival models.

Dominion versus Stewardship

There is an interesting parallel to the youth–maturity theme in the Scriptures. Early on in the Bible we are told to "be fruitful, and multiply" and take dominion over the earth (Genesis 1:28). One interpretation of this message is that it was meant to apply to all living things, not just humans (Bratton 1992). Elsewhere in the Scriptures (Luke 12:48; I Corinthians; Psalm 37:27–28, among others) we are admonished to become "stewards" (a word that comes from roots meaning "keeper of the house") and to take care of the earth. A reasoned interpretation is that these messages are not contradictory, nor a matter of right or wrong, but constitute a sequence in time; that is, there are times and places for each.

In the early development of civilizations, taking dominion over the environment and exploiting resources (clearing the land for crops, mining the earth for materials and energy, etc.) and high birth rates are both necessary for human survival. But as our society becomes ever more crowded, resource-demanding, and technologically complex, there is less need for large families and child labor, and more important, various limitations are reached that force us to turn to stewardship in order not to destroy our life-supporting house. As discussed in Chapter 6, this stage is sometimes called the **demographic transition**.

There are also youth–maturity parallels in the development of the individual and of the ecosystem. The transition in human development is called adolescence—a difficult time that we all experience. The transition in ecosystem development is called ecological succession, covered in Chapter 7. Many world leaders are beginning at least to talk about parallel transitions in economic development, as we will see in the next section.

Economic Growth versus Economic Development

In 1987, a World Commission on Environment and Development issued a report entitled *Our Common Future*, which has come to be known as the "Brundtland Report" after Madame Brundtland, the former Prime Minister of Norway and chair of the commission. The report concludes that the current trends of economic development and accompanying environmental degradation are unsustainable. Irrevocable damage to planetary ecosystems is suppressing the economic status of much of the world's population. Survival depends on *changes now*. The first step in bringing about changes is to seek ways to enhance cooperation between nations so that they can work together toward global sustainability. The report is important not so much for what it says as for the fact that a group of 23 political leaders and scientists from both rich and poor countries could agree that the health of the global environment is important for everyone's future.

In 1991 UNESCO issued a report entitled *Environmentally Sustainable Economic Development: Building on Brundtland* (Goodland et al. 1991). This report makes a distinction between economic *growth*, which involves getting larger (quantitative growth), and economic *development*, which involves getting better (qualitative growth) without increasing the total consumption of energy and materials beyond a level that is reasonably sustainable. The report concludes that "a five-to-tenfold expansion of anything remotely resembling the present economy [which some economists say is necessary to reduce poverty worldwide] would simply speed us from today's long-run unsustainability to imminent collapse." Therefore the economic growth required for poverty reduction (especially in the less developed countries) "must be balanced by negative throughput growth for the rich."

In 1992 world leaders convened an **Earth Summit** in Rio de Janeiro, Brazil, in search of international agreements that could help save the world from pollution, poverty, and the waste of resources. Confrontation between the wealthy "north" and the poor "south" dominated the proceedings, and few meaningful agreements were reached; however, the concept of sustainable development did emerge as a means of combining economic and ecological needs. Many who attended the summit came away with the feeling that a pathway had been opened for future cooperation among nations.

The human disposition being what it is—i.e., we wait until a problem gets *really* bad before taking action—it often takes a crisis or a disaster to bring about good environmental planning and a start on the transitions we have been discussing. The following example describes how a local disaster was followed by economic development without an increase in the size of the city (Flanagan 1988):

> In 1972 Rapid City, South Dakota, suffered a devastating flash flood on Rapid Creek that damaged $160 million in property, destroyed 1,200 buildings, and killed 238 people. Through the leadership of

the mayor, Don Barnett, the community instituted a national proto-type floodplain acquisition program, removed the damaged homes from the floodplain, and created a six-mile long, quarter-mile wide urban greenway through the center of the city. The greenway now contains parks, recreation trails and golf courses. Rapid Creek was stocked with sport fish and is now the most popular recreational fishing stream in the entire state. Rapid City stands as a creative example of enlightened leadership, turning a disaster into a multi-use community asset that benefits all aspects of the city, including commerce and the tourist industry.

The Human Landscape: Setting Priorities

Because human predicaments differ dramatically in different parts of the world, priorities for seeking solutions must differ accordingly. In terms of current economic conditions and population densities, William Clark (1989) divides the world into the following four regions:

1. Low-income, high-density regions (Examples: India, Mexico)
2. Low-income, low-density regions (Examples: Amazonia, Malaya/Borneo)
3. High-income, low-density regions (Examples: United States, Canada, the oil-rich desert kingdoms)
4. High-income, high-density regions (Examples: Japan, northwestern Europe)

In low-income countries the first priority is reducing poverty, which means promoting sustainable economic growth. In high-density countries reducing birth rates by family planning or other means should be paramount. Reducing waste and overconsumption of resources should be the first priority of high-income countries, which means shifting from quantitative to qualitative growth (true economic development), as discussed in the previous section. Most high-income countries are already well into the demographic transition (i.e., reduction of their population growth rates).

The Paradox of Technological Development

Just about every technological advance that is intended to improve our well-being and prosperity has its dark as well as its bright side. Or, as Paul Gray, engineer and president of M.I.T., states (1992), "A paradox of our time is the mixed blessing of almost every technological development." We have described a number of examples of this paradox in previous chapters, including the mixed blessings of plant pest and disease control technology and Green Revolution technology. The bright side of these technologies is increased yield of food produced with less labor. The dark side is the heavy use of fertilizers, pesticides, and machinery, which results in widespread air, water, and soil pollution, the development of resistant strains of pests, and serious rural unemployment. Another example is the coal-fired power plants that provide electricity for most of the United States, but also make a major contribution to acid rain.

The point to be made here is that as we seek new technologies, we must be aware that they will have dark sides that must not only be anticipated but also dealt with. Often what is needed is a **counter-technology** that will at least ameliorate the detrimental effects. In agriculture, conservation tillage is a counter-technology that is being widely adopted (see Chapter 4 and Figure 15 in Chapter 5). In the case of power plants, "clean coal" technology that would eliminate acid emissions is available (Spencer et al. 1986).

Like the evolutionary arms race (see Chapter 7), action–counter-action is the way of life in the affairs of humans as in nature. In nature, natural selection, when allowed to run its course, takes care of the counter-action, but in human affairs we have to apply our own negative feedback. The danger is either in not anticipating the need for counter-action or in waiting too long to act. The delay in implementing clean coal technology, for example, is due to the fact that there is a **transition cost** in making the change. Governments should assist utilities, and industry in general, in making the transition by providing tax relief, grants, or other incentives rather than promoting the status quo ("business as usual")—as governments are wont to do unless you, the public, tell politicians you want changes that will improve health and the quality of life, even if there is a temporary economic cost. As we have emphasized many times, anything that improves the quality of our environment and our lives benefits the economy in the long run.

Restoration Ecology Revisited

Because so much of the environment has been damaged beyond nature's ability to repair it, restoring damaged ecosystems is becoming big business, as noted in the section entitled "When Succession Fails" (Chapter 7). Ways and means of restoring wetlands that were drained or destroyed before their value as life-support buffers was recognized is an especially active area of research. A pioneer in developing the new field of **restora-**

tion ecology is John Cairns, Jr. Since 1971 he has written and edited several large books on the subject. He presents a number of case studies in a 1992 report that he organized for the National Academy of Sciences. In reviewing such environmental engineering projects, it is evident that they are most successful when four key groups work together in a coordinated manner—namely, citizens' groups, governmental agencies (local, state, and federal), science and technology, and business interests. When any one of these groups is not strongly involved, restoration projects usually fail to achieve their long-term goals. (For more on environmental engineering see Mitsch and Jorgensen 1989.)

Input Management Revisited

The strategy of managing inputs rather than outputs was first mentioned in Chapter 1 as an "about-face" that is necessary as a means of reducing pollution. **Input management** of production systems (e.g., agriculture, power plants, and manufacturing) is a practical and economically feasible approach to improving and sustaining the quality of our life-support systems (Odum 1989). This concept is illustrated in Figure 1. As shown in Figure 1A, attention in the past has focused on increasing outputs, that is, yields, by pouring in resources (such as fertilizers and fossil fuels) without much regard to efficiency or the amount of unwanted outputs (such

The Two "C's" for Success in Environmental Protection

In discussing the Chesapeake Bay in Chapter 1, we noted that there was now a public **consensus** that action was needed to save the bay, and that efforts had begun to organize a **coalition** of diverse interests to get the job done. The interests that must work together include (1) active public or private organizations (Friends of the Bay, for example), (2) governments, especially local and state, (3) business interests, which must recognize that maintaining environmental health is good business in the long run, and (4) educational and research institutions that can provide the needed science and technology. None of these special interest groups can accomplish anything on a large scale working alone, as was emphasized in the discussion of restoration ecology in this chapter.

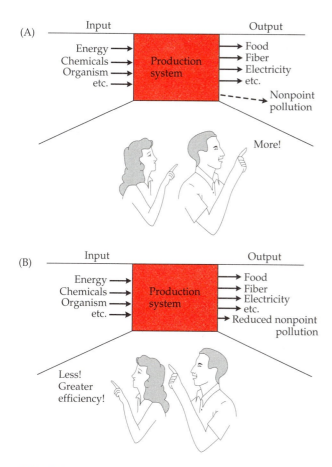

FIGURE 1
The "about-face" needed on management of production systems. (A) Focus on output, such as yield, with the consequences of increased non-point-source pollution. (B) The shift to input management, with focus on efficiency and reduction of costly and environmentally damaging inputs so as to reduce nonpoint-source pollution. (After Odum 1987).

as nonpoint pollution) created. Input management involves an about-face, as shown in Figure 1B, with the goal of reducing inputs to only those that can be efficiently converted to the desired product. Input management can also be called **top-down management** since it involves assessing inputs to the whole system (as the external forcing functions) *first*, then internal dynamics and outputs *second*. Applying the concept to wastes means that *waste reduction takes precedence over waste disposal*. As noted in

Chapter 1, there is no way New York City can continue to dispose of its wastes in the ocean without reducing their volume.

Professor Luo Shi Ming of South China Agricultural University, at a seminar at the University of Georgia in 1988, suggested that the proper path for Third World countries to take in developing their agriculture is to bypass the wasteful, high-input stage and go directly from their traditional agriculture to new low-input practices, using strategies such as genetic biotechnology, which can create plants that require less energy subsidy and environment-damaging chemicals. Why not the same strategy for industrial development?

Human Behavior and Environmental Change

As we noted earlier in this Epilogue, humans tend to wait until things get really bad before taking rational corrective action. It is hard to understand why this is so (Holden 1988; Stern 1993). As emphasized by E. O. Wilson in his writings on sociobiology (Wilson 1975), humans and animals share a common genetic basis for behavior. Thus, we tend to react to stress by becoming violent, perhaps because survival in the past required that we develop weapons to confront a hostile world and competing tribes. Sociologist W. R. Catton (1976) worries about this human tendency to become "beastly" when crowded. He suggests that this tendency might be ameliorated if people understood what changes are needed to reduce stress, which, of course, is what this Epilogue is all about.

Environmental Ethics and Aesthetics

Maintaining and improving environmental quality requires an ethical underpinning. *Not only must it be against the law to abuse nature's life-support systems, it must be understood to be unethical as well.* One of the most widely read and cited essays on the subject of **environmental ethics** is Aldo Leopold's "The Land Ethic," first published in 1933 and included in his classic book, *A Sand County Almanac* (1949). Leopold spent his early years as a forester in a part of the West that could be reached only on horseback, and where the howl of the wolf could still be heard. Later he pioneered the field of game management and became a professor. He and his family spent as much time as possible at a cabin (which has since become a shrine for conservationists) on an old worn-out farm they bought and restored as a place of natural beauty in Sauk County, Wisconsin. Aldo Leopold will be best remembered for the writing he did there— writing of the caliber of Henry David Thoreau's (New England's spokesman for the wild and the beautiful).

Leopold opens "The Land Ethic" by describing how Odysseus, hero of *The Odyssey*, on returning from the wars in Troy hanged a dozen slave girls whom he suspected of misbehavior during his absence. "This hanging involved no question of propriety. The girls were property. The disposal of property was then, as now, a matter of expediency, not of right or wrong." The concepts of right and wrong were not lacking in ancient Greece, but they did not extend to slaves. Since then, of course, human rights have received increasing ethical as well as legal and political attention. But what about the rights of other organisms and the environment? Leopold defined an **ethic** ecologically as "a limitation on freedom of action in the struggle for existence," and philosophically as "a differentiation of social from anti-social conduct." He suggested that the extension of ethics over time is a sequence, as follows: First, there is the development of religion as a human-to-human ethic. Then comes democracy as a human-to-society ethic. And finally, there is a *yet to be developed* ethical relationship between humans and their environment; in Leopold's words, "the land-relation is still strictly economic, entailing privileges but not obligations."

As we have attempted to document in this book, there are strong scientific and technical reasons for the proposition that an expansion of ethics to include the life-support environment is necessary for human survival. There are many legal mechanisms, such as **conservation easements**, available that can encourage landowners to trade their options to develop property for tax relief or other favorable economic considerations. It is also encouraging that in the past decade there has been a great increase in the number of articles, books, college courses, and journals that deal with environmental ethics (Rolston 1986; Callicott 1987; Potter 1988; Hargrove 1989; Ferre and Hartel 1994).

Dual Capitalism

A recurrent theme in this book has been the contention that the overly narrow economic theories and policies that dominate world markets and world politics are major obstacles to achieving a reasonable, common-sense balance between our need for nonmarket and market goods and services. Around the turn of the twentieth century, a group of scholars calling themselves "holistic economists" formed a school critical of economic models of the day (Grundy 1943). Efforts to establish a holistic economics at that time, however, were drowned, as it were, in the flood of oil that spawned rapid growth in monetary and material wealth. Classic growth theory served well as long as the supply of cheap oil far exceeded demand. As the oil age is now peaking and global pollution and overshoots are rampant, it seems that the time has come to redevelop some kind of **holoeconomics** that includes cultural and environmental values along with monetary ones—in other words, an economics that gives equal consideration to market capital and natural capital (Daly 1990).

It should be possible with appropriate regulatory–incentive measures to develop such a **dual capitalism** gradually over a period of time (Odum 1997). Under such a system, a business or industry would not only consider the market possibilities for a new product or service, but also plan how to produce the product or service with efficient use of resources and with as much recycling and as little pollution as possible (there are lots of opportunities for new technology here!). It would also consider how to internalize source reduction and waste management costs so that the consumer, rather than the taxpayer, pays for waste management.

It is encouraging to see that, beginning with the first international conference on the integration of economy and ecology in 1982 (Jansson 1984), the economic–ecology interface has been receiving increasing attention. For a review, the book of essays edited by Daly and Townsend (1993) is recommended.

A Host-Parasite Model for Humans and the Biosphere

As pointed out in Chapter 1 and again in Chapter 6, humanity is in reality a parasite on the biosphere. As discussed in some detail in Chapter 6, ecological research on parasite–host relationships is revealing that coevolutionary natural selection operates to promote reciprocal adaptations so that the parasite does not eliminate its host, and thereby itself. Given time and freedom from too much outside disturbance, hosts develop resistance, and parasites reduce their virulence and establish "reward feedbacks" that *benefit* the host. Figure 2 is a survival model depicted as a host–parasite system.

If we are to be a prudent parasite that survives, we must devote more attention to reducing our wastes, resource destruction, and overpopulation in order to reduce our virulence. Equally important, we must invest more in earth care (the service/reward feedback loop in Figure 2). Otherwise we will become a malignant cancer that will ultimately destroy our host.

An Ethics Survival Model

Alternate **scenarios** (a "scenario" is an outline of a sequence of scenes or events) that may determine the quality of future survival for humanity

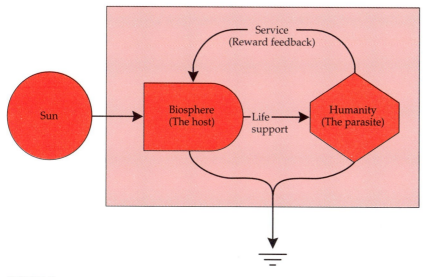

FIGURE 2

A life-support model shown as a host–parasite system. As a prudent parasite, humanity must service (i.e., preserve, maintain, and repair) the biosphere if we expect to continue to receive high-quality life-support goods and services.

are shown in Figure 3. These scenarios are not predictions, since, as we have already stressed, no one (and no computer) can really predict the future; they are more like weather forecasts that have a certain probability of being right or wrong.

The left-hand sequence starts with the assumption that we will continue to take the short-term view and restrict ethics and law to protecting and promoting the welfare of the individual (i.e., with little regard for the public welfare, assuming that what's good for the individual is always good for society and the world). The logical consequences of placing value only on the individual are continued rapid expansion of world population and stressed and degraded life-support ecosystems. Together these will lead to a less than satisfactory life for all but perhaps a few very rich people, since air, food, and water will be increasingly poor in quality and short in supply.

The alternate scenario (the right-hand sequence) is based on the assumption that we will turn more and more to the long-term view, with value placed on species (ours and all the others) and on maintaining healthy ecosystems worldwide. The logical consequences of extending

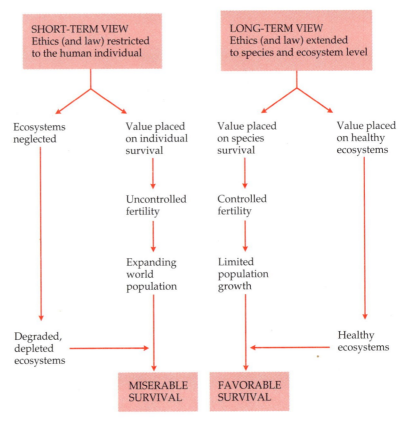

FIGURE 3
A survival model of two contrasting scenarios. (After Potter 1988.)

ethics and law to the species and ecosystem levels are reduced population growth (with stabilization in the next century) and healthy life-support systems, leading to favorable survival for all people and for all life.

Summary: Transitions over the Long Term

A summary of youth-to-maturity parallels between systems and long-term prerequisites for achieving maturity is given in Figure 4. If human society can make these transitions, then we can be optimistic about the future of humankind. To do this, we must merge the "study of the house-hold" (ecology) and the "management of the household" (economics),

Youth-to-Maturity Parallels
Individual: Transition is called "adolescence"
Biotic community: Transition is called "ecological succession"
Society: Transition is called "demographic transition"
Energy flow shift: In early development , the major flow of energy must be directed to growth. In later development, an increasing proportion of available energy must be directed to maintenance and control of disorder **(see Figure 3 in Chapter 7)**.
Competition–Cooperation: In nature and society, cooperation pays when systems get complex and tight (resources are limiting).
Network "Law": The cost of maintenance (C) increases as a power function—roughly, as a square—of the number of network services (N). As a city doubles in size, cost of maintenance more than doubles **(see Chapter 7)**.
Prerequisites for Maturity (Sustainability)
1. Market capitalism to dual capitalism
2. Boom-and-bust growth to sigmoid growth (or quantitative to qualitative growth)
3. Haphazard development to regional land use planning
4. Competition to mutualism (or confrontation to cooperation)
The Ultimate Measure
$P \geq R$
Biotic community: Production $\geq$ respiration
Population: Natality $\geq$ mortality
Society: Production $\geq$ maintenance
When **R** exceeds **P** the system cannot be sustained

FIGURE 4
Youth-to-maturity parallels between individual, biotic community, and
society, and prerequisites for achieving maturity.

and our ethics must be extended to include environmental as well as
human values. Accordingly, bringing together these three "E's" is the ultimate holism and the great challenge for our future. To bring about the
needed changes and reforms, we need to add the two "C's," consensus
and coalition. And, finally, if we can "dualize" our current capitalism, we
can really be optimistic about our future.

Suggested Readings

Ambio. 1992. Special issue: Environment and Biodiversity Loss. Vol. 21(3). (A journal of the human environment, published by the Royal Swedish Academy, Stockholm.)

*Ausubel, J. H. 1996. Can technology spare the earth? *Am. Sci.* 84: 166–178. (See also "The Liberation of the Environment," special issue of *Daedalus*, vol. 125, no. 3, 1996.)

*Bernstein, B. B. 1981. Ecology and economics: Complex systems in changing environments. *Annu. Rev. Ecol. Syst.* 12: 309–330.

*Bratton, S. 1992. *Six Billion and More: Human Population Regulation and Christian Ethics*. Westminster/John Knox, Louisville, KY.

*Brown, L. R., ed. *The State of the World*. Worldwatch Institute, Washington, D.C. (Annual volumes, the first published in 1984, reviewing the status of environment and resources.)

*Butzer, K. W. 1980. Civilizations: Organisms or systems? *Am. Sci.* 68: 517–523.

*Cairns, J. 1988. *Rehabilitating Damaged Ecosystems*. CRC Press, Boca Raton, FL.

*Cairns, J. 1992. *Restoration of Aquatic Ecosystems*. National Academy Press, Washington, D.C.

*Callicott, J. B., ed. 1987. *Companion to A Sand County Almanac*. University of Wisconsin Press, Madison.

*Capra, F. 1982. *The Turning Point*. Bantam Books, New York.

*Carson, R. 1962. *Silent Spring*. Houghton Mifflin, Boston.

Catton, W. R. 1976. Can irrupting man remain human? *BioScience* 26: 262–267. (Reprinted in *Carrying Capacity Network: Focus* 3(2): 19–25, 1993.)

*Catton, W. R. 1980. *Overshoot*. University of Illinois Press, Urbana.

*Clark, W. C. 1989. Managing planet Earth. *Sci. Am.* 261(3): 47–54.

Conrad, M. 1983. *Adaptability: The Significance of Variability from Molecule to Ecosystem*. Plenum Press, New York.

*Costanza, R. 1987. Social traps and environmental policy. *BioScience* 37: 407–412.

Costanza, R., ed. 1991. *Ecological Economics: The Science and Management of Sustainability*. Columbia University Press, New York.

Costanza, R., B. Norton, and B. Haskell, eds. 1992. *Ecosystem Health: New Goals for Environmental Management*. Island Press, Covelo, CA.

*Cross, J. G., and M. J. Guyer. 1980. *Social Traps*. University of Michigan Press, Ann Arbor.

*Daly, H. 1973. *Toward a Steady-State Economy*. W. H. Freeman, San Francisco.

*Daly, H. 1989. Economics, environment, and community. *Earth Ethics* 1(1): 9–11. (Emphasizes the difference between quantitative and qualitative growth.)

Daly, H. 1990. Towards some operational principles of sustainable development. *Ecol. Econ.* 2: 1–6.

*Daly, H., and J. B. Cobb. 1989. *For the Common Good: Redirecting the Economy Toward Community for the Environment and a Sustainable Future*. Beacon Press, Boston.

Daly, H., and K. N. Townsend. 1992. *Valuing the Earth: Economics, Ecology, Ethics*, 2nd ed. M.I.T. Press, Cambridge, MA. (Fourteen essays, several of which have become classics.)

Eckholm, E. P. 1982. *Down to Earth: Environment and Human Needs*. W. W. Norton, New York. (Report prepared in commemoration of tenth anniversary of the historic Stockholm conference on the human environment.)

*Edney, J. J., and C. Harper. 1978. The effect of information in resource management: A social trap. *Human Ecol.* 6: 387–395.

*Ehrlich, P. R. 1968. *The Population Bomb*. Ballantine Books, New York.

Ehrlich, P. R., and A. H. Ehrlich. 1990. *The Population Explosion*. Simon & Schuster, New York. (A sequel to *The Population Bomb*.)

Ferre, F., and P. Hartel, eds. 1994. *Ethics and Environmental Policy*. University of Georgia Press, Athens.

*Flanagan, R. D. 1988. Planning for multi-purpose use of greenway corridors. *Natl. Wetlands Newsletter* 10(2): 7–8. (Published by the Environmental Law Institute, Washington, D.C.)

Gilliland, M. W., ed. 1978. *Energy Analysis: A New Public Policy Tool*. AAAS Selected Symposium, no. 9. Westview Press, Boulder, CO.

Gore, A. 1992. *Earth in the Balance: Ecology and the Human Spirit*. Houghton Mifflin, Boston. (The vice president proves he is environmentally literate and understands the tough political choices that must be made.)

*Gray, P. E. 1989. The paradox of technological development. In *Technology and Environment*, pp. 192–205. National Academy Press, Washington, D.C.

*Goodland, R., H. Daly, S. E. Serafy, and B. von Droste, eds. 1991. *Environmentally Sustainable Economic Development: Building on Brundtland*. United National Education, Scientific, and Cultural Organization (UNESCO), Paris.

*Grundy, A. G. 1943. *Modern Economic Thought: The American Contribution*. Prentice-Hall, Englewood Cliffs, NJ.

*Hardin, G. 1968. The tragedy of the commons. *Science* 162: 1243–1248.

*Hardin, G. 1985. *Filters against Folly*. Viking Press, New York.

*Hargrove, E. C. 1989. *Foundations of Environmental Ethics*. Prentice-Hall, Englewood Cliffs, NJ.

Hawkins, P., J. Ogilvy, and P. Schwartz. 1982. *Seven Tomorrows: Toward a Voluntary History*. Bantam Books, New York. (Stanford "think tank" argues that nations and people move ahead when there is a common vision that motivates. They forecast the development of a "transformational alternative" that takes the best from both the "left" and the "right," thus combining the individual and the public good.)

Holden, C. 1988. The ecosystem and human behavior. *Science* 242: 663. (Report on a workshop at which social scientists discussed human behavior and global change.)

*Jansson, A.-M., ed. 1984. *Integration of Economy and Ecology: An Outlook for the Eighties*. Proc. Wallenberg Symposia, Stockholm.

*Kahn, A. E. 1966. The tyranny of small decisions: Market failures, imperfections and the limit of economics. *Kyklos* 19: 23–47.

*Kahn, H., W. Brown, and L. Martel. 1976. *The Next 200 Years*. William Morrow, New York.

*Laszlo, E. 1977. The Club of Rome of the future vs. the future of the Club of Rome. In *Goals in a Global Community*, eds. E. Laszlo and J. Bierman. Pergamon Press, New York.

*Leopold, A. 1949. The land ethic. *A Sand County Almanac*. Oxford University Press, New York. (See also earlier version in *J. For.* 31: 634–643, 1933.)

*Marsh, G. P. 1864. *Man and Nature, or Physical Geography as Modified by Human Nature*. 1965 reprint edited by D. Lowenthal. Harvard University Press, Cambridge, MA. (For an evaluation of Marsh's classic, see F. Russell in Horizon 10: 17–23, 1968.)

*McHarg, I. L. 1992. *Design with Nature*. John Wiley, New York. (Revised edition of a book first published in 1969. McHarg pioneered the map-overlay method of land-use planning in metropolitan districts.)

*McNamara, R. S. 1984. Time bomb or myth: The population problem. *For. Aff.* 62: 1107–1131.

McPhee, J. 1989. *The Control of Nature*. Farrar, Straus & Giroux, New York. (McPhee, an award-winning journalist, has written more than 20 books on human encounters with the forces of nature. Here he explores the engineering mindset bent on controlling the Mississippi River, volcanic flows in Iceland, and mud slides in California.

*Meadows, D. H., D. L. Meadows, J. Randers, and W. W. Behrens. 1972. *The Limits to Growth: A Report for the Club of Rome's Project on the Predicament of Mankind*. Universe Books, New York. (This is the book that started worldwide controversy on the future of growth economics.)

*Meadows, D. H., J. Richardson, and C. Bruckmann. 1982. *Groping in the Dark: The First Decade of Global Modelling*. John Wiley, New York.

*Meadows, D. H., D. L. Meadows, and J. Randers. 1992. *Beyond the Limits: Confronting Global Collapse, Envisioning a Sustainable Future*. Chelsea Green, Post Mills, VT.

*Mitsch, W. J., and S. E. Jorgensen. 1989. *Ecological Engineering: An Introduction to Ecotechnology*. John Wiley, New York.

Moore, J. W. 1986. *The Changing Environment*. Springer-Verlag, New York. (Reviews principal environmental issues of the day in both industrial and developing nations.)

*Morehouse, W., and J. Sigurdson. 1977. Science, technology and poverty. *Bull. Atom. Sci.* 33: 21–28.

*Nader, L., and S. Beckerman. 1978. Energy as it relates to the quality and style of life. *Annu. Rev. Energy* 3: 1–28.

Nash, R. 1982. *Wilderness and the American Mind*, 3rd Ed. Yale University Press, New Haven, CT. (In the Epilogue, Nash contrasts a future dominated by huge cities and urban sprawl and a completely domesticated "garden world" of dispersed rural development and small cities. In both cases there will be no wilderness or other completely natural areas unless they are preserved by positive "zoning" with some kind of licensing system to regulate the number of users.)

*National Academy of Sciences. 1971. *Rapid Population Growth*. Johns Hopkins Press, Baltimore.

*National Academy of Sciences/The Royal Society of London. 1992. *Population Growth, Resource Consumption, and a Sustainable World*. Joint statement.

*Odum, E. P. 1977. Ecology—the common sense approach. *The Ecologist* 7: 250–253.

Odum, E. P. 1983. Epilog. *Basic Ecology*. Saunders College Publishing, Philadelphia.

*Odum, E. P. 1989. Input management of production systems. *Science* 243: 177–182. (See also *J. Soil Water Conserv.* 42: 412–414, 1982.)

Odum, E. P. 1997. Source reduction, input management, and dual capitalism. (In press.)

*Odum, W. E. 1982. Environmental degradation and the tyranny of small decisions. *BioScience* 32: 728–729.

*Office of Technology Assessment, U.S. Congress. 1982. *Global Models, World Futures, and Public Policy*. U.S. Government Printing Office, Washington, D.C.

Orr, D. W. 1992. *Ecological Literacy: Education and the Transition to a Postmodern World*. State University of New York Press, Ithaca.

*Osborn, F. 1948. *Our Plundered Planet*. Little, Brown, Boston.

*Platt, J. 1973. Social traps. *Am. Psychol.* 28: 641–651.

*Potter, V. R. 1988. *Global Bioethics: Building on the Leopold Legacy*. Michigan State University Press, East Lansing. (See also: *Persp. Biol. Med.* 30: 157–169, 1987.)

Rapport, D. J. 1992. Evaluating ecosystem health. *J. Aquat. Ecosyst.* 1: 15–24.

Renner, M. 1991. *Jobs in a Sustainable Economy*. Worldwatch Paper no. 104. Worldwatch Institute, Washington, D.C. (Jobs lost in extractive and polluting industries can be replaced with new jobs in service, recycling, pollution control, and renewable energy industries. The government should subsidize the transition costs.)

*Rolston, H. 1986. *Philosophy Gone Wild: Essays in Environmental Ethics*. Prometheus Books, Buffalo, NY.

Schmidheiny, S. 1992. *Changing Course: A Global Perspective on Development and the Environment*. M.I.T. Press, Cambridge, MA. (Examples of how business is changing to meet the challenges of the global environment.)

Schneider, S. H., and L. Morton. 1981. *The Primordial Bond: Exploring Connection between Man and Nature through the Humanities and Sciences*. Plenum Press, New York.

*Schumacher, E. F. 1973. *Small Is Beautiful: Economics As if People Mattered*. Harper & Row, New York. (But unfortunately big is more powerful!)

*Sears, P. 1935. *Deserts on the March*. University of Oklahoma Press, Norman.

*Seligson, M. A. 1984. *The Gap between Rich and Poor: Contending Perspectives on Political Economy and Development*. Westview Press, Boulder, CO. (Between 1950 and 1980, the per capita income gap between rich and poor nations grew from $3677 to $9648. The gap is also widening within rich nations.)

*Simon, J. L. 1981. *The Ultimate Resource*. Princeton University Press, Princeton, NJ. (Human ingenuity can overcome any resource shortage!)

Simon, J. L., and H. Kahn, eds. 1984. *The Resourceful Earth: A Response to Global 2000*. Blackwell, New York.

Spencer, D. F., A. S. Alpert, and H. H. Gilman. 1986. Cool water: Demonstration of a clean and efficient new coal technology. *Science* 232: 609–612.

Speth, J. G. 1984. *The Global Possible: Resources, Development and the New Century*. World Resources Institute, Washington, D.C.

Stern, P. C. 1993. A second environmental science: Human-environment interactions. *Science* 260: 1897–1899.

Teitenberg, T. H. 1991. Managing the transition: The potential role of economic policies. In *Preserving the Global Environment*, ed. J. T. Mathews. Norton, New York.

*Toynbee, A. J. 1961. *A Study of History*. Oxford University Press, New York.

Villee, C. A., ed. 1986. *Fallout from the Population Explosion*. Paragon House, New York. (Compilation of contemporary writings that avoid extremes of hysteria and complacency.)

Vitousek, P. M. 1992. Global environmental change: An introduction. *Annu. Rev. Ecol. Syst.* 23: 1–14. (Commentaries on changes in the atmosphere, climate, ozone, land use, biodiversity, and biological invasions.)

*Vogt, W. 1948. *The Road to Survival*. Sloane, New York.

*Watt, K. E. F., L. F. Molloy, C. K. Varshney, D. Weeks, and S. Wirosardjono. 1977. *The Unsteady State: Environmental Problems, Growth, and Culture*. University Press of Hawaii, Honolulu.

*Wilson, E. O. 1975. *Sociobiology.* Harvard University Press, Cambridge, MA.
*World Commission on Environment and Development. 1987. *Our Common Future.* Oxford University Press, New York.

*Indicates references cited in this chapter

Index

ABOUT THE BOOK

Editor: Andrew D. Sinauer
Project Editor: Carol J. Wigg
Copy Editor: Norma Roche
Production Manager: Christopher Small
Book Production: Maggie Haddad
Art: Precision Graphics
Book Design: Joseph Vesely and Christopher Small
Cover Design: Jefferson C. Johnson
Book and Cover Manufacture: Best Book Manufacturers